Frontiers of Space

Herbert Friedman, General Editor

INTERNATIONAL COOPERATION IN SPACE

The Example of the European Space Agency

Roger M. Bonnet and Vittorio Manno

Harvard University Press · Cambridge, Massachusetts · London, England · 1994

Library of Congress Cataloging-in-Publication Data

Bonnet, R.-M. (Roger-M.)
    International cooperation in space : the example of the European
Space Agency / Roger M. Bonnet and Vittorio Manno.
        p.   cm. — (Frontiers of space)
    Includes bibliographical references and index.
    ISBN 0-674-45835-4 (acid-free)
    1. Astronautics—International cooperation.   2. European Space
Agency—History.   I. Manno, Vittorio.   II. Title.   III. Series.
TL788.4.B66   1994
387.8′0604—dc20   94-16206
                                                                          CIP

To Ariane, Domitilla, and Niccolò

ACKNOWLEDGMENTS

We wish to express our most sincere appreciation of Herb Friedman's invitation to undertake the writing of this book. This has been a challenge for both of us at a time of major changes on the world political scene, in Europe in particular, and at times of crucial decisions to be made concerning ESA's long-term plan.

We are grateful to the two Directors General of ESA, Reimar Lüst and Jean-Marie Luton, for having authorized this undertaking. This work would have been much more difficult and certainly less complete without the help and assistance from many people in ESA. Our acknowledgement goes therefore to: Jean Arets, Jean-Louis Boucher, Martin Brady, Giacomo Cavallo, Herbert Frank, Giulio Grilli, Gabriel Lafferranderie, Marie-Angèle Lemoine, René Oosterlinck, Arne Pedersen, Jean-Pierre Provost, Ian Pryke, Winfried Thoma, Georges van Reeth, Simon Vermeer, and Peter Wenzel.

We would like also to thank Rachel Villain from Euroconsult for having authorized the reproduction of Figure 20, from their excellent *World Space Industry Survey*.

The help and support of Dina Bauer and Valérie Lécuraud are deeply appreciated.

The opinions expressed in the book reflect the viewpoint of the authors and not necessarily those of the organizations to which they belong.

CONTENTS

Introduction   1

1. The Birth of ESA   3

2. Governing Principles   24

3. The Agency and Its Member States   60

4. International Connections   73

5. Two Cases of International Cooperation   98

6. The New Context   120

Conclusion   136

Notes   141

Bibliography of Works in English   145

ESRO and ESA Spacecraft   147

The Horizon 2000 Long-Term Plan   151

List of Illustrations   154

Index   155

International Cooperation in Space

INTRODUCTION

This book describes a special case in the development of international cooperation in space. The creation of organizations such as the European Space Research Organization (ESRO) and the European Space Agency (ESA), involving in friendly and peaceful cooperation nations which twice in the century had been at war with one another, sets an example of a unique character. This example is particularly relevant to today's political situation when another war, the "cold war" between the West and the East, seems to be behind us. This international development opens an opportunity for a definition of a new framework of cooperation in space. This is even more necessary in a context where the space race has slowed down and the need to combine intellectual and financial resources is becoming more pressing because of the increasing size and scope of space programs.

For more than thirty years now the Europeans have been successfully cooperating in the framework of ESRO and ESA. At the same time, they have deserted neither their own domestic programs nor their bilateral or multilateral cooperative agreements with other agencies in Europe or elsewhere in the world. This mosaic of activities is characteristic of Europe and reflects its differentiated political structure and economic capabilities. It is a very successful example of scientific and technical cooperation involving several nations determined to join forces with a common goal in mind.

Within the overall scope of space activities, space science has, from very early days, offered one of the best possible examples of fruitful and peaceful international cooperation. This is mainly because it rests on a policy of an open exchange of information, of data, and of all results obtained with the missions. It is also because the definition and selection of science projects has been protected from considerations of a nonscientific nature that would have conflicted with the fundamental principle of a search for scientific excellence. Therefore, it was natural that the first European international space organization should have purely scientific and peaceful goals.

Today, in the 1990s, fundamental science (excluding Earth observation and microgravity), represents only about 13 percent of the Agency's overall budget, equivalent to 330 million accounting units (AU) in 1993 economic conditions, or about $390 million. The rest of the budget (2.5 billion AU or about $3 billion) covers the development of the Ariane launcher, the participation in the international space station, telecommunications, Earth observations, and microgravity programs. In spite of this relatively small percentage, space science in ESA has a very central and privileged position. The Science Program is the only mandatory program and, as such, is the only program on which the cohesion of ESA rests.

As scientists who are or have been responsible for the management of that program at ESA, we feel naturally inclined toward using space science as the best example of multinational cooperation in Europe. We have deliberately not discussed the commercial activities which involve the exploitation of the Ariane launcher, the distribution of Earth observation data, or the exploitation of telecommunications satellites. We have also abstained from discussing the manned space program. We do discuss the aspects of these programs when they are of direct relevance to international cooperation; in particular, we dwell on the case of the space station because of its great importance to the future of global cooperation.

During the period covering the writing of this book, 1991–1994, the face and the political structure of the world have changed at an incredibly fast rate. We have modified the content of the book wherever necessary to take these events into account.

1

THE BIRTH OF ESA

In the aftermath of the Second World War and of the early success of the Soviet and American space programs, it became clear to a few European scientists that an international space research organization was needed in Europe. Only such an organization would allow Europe to speak with one voice to the two leading space powers and at the same time to build a competitive space program. It would also help prevent a brain drain of the best European scientists to the United States. The alternatives would have been for European nations to compete individually or to participate in an organization under the auspices of NATO.[1] Because of the military character of NATO and the desire expressed by several European scientists to keep open the possibility of dialogue with their Soviet colleagues, this option was rejected, making it implicit that the future European Space Research Organization would have exclusively peaceful objectives.

A more pragmatic motivation was the early realization that space activities might lead to advances in technology that could be important in the resurrection of Europe's economic and industrial development. The lack of a European military space program at a level comparable with that of the Soviets or of the Americans made it difficult for the European aerospace industry to reach critical levels of technological know-how and launching capability. A civilian space program offered a possible remedy to this state of affairs.

The need for a strong European entity as an essential element of the global political balance explains both why Europe has been the prime beneficiary of the internationalization of the American space program and why the push for European autonomy has grown stronger as Europe has gained momentum as well as scientific and industrial experience. At the outset, having started very late in the game, Europe had few options other than to cooperate with the two space superpowers.

Considerations of competitiveness and self-sufficiency have played an important role in European space policy. The European Space Agency (ESA), which succeeded the European Space Research Organization (ESRO) in 1975, is an inherently international agency whose aims are to cooperate, for exclusively peaceful purposes, in developing scientific and application satellites with a view to building a competitive European industrial capability. The very successful development of these satellites, as well as of Spacelab and the Ariane launcher, proves that these aims have been reached. In this context, it should not come as a surprise that industrial policy still plays a major role in the overall running of the agency and of its programs. We shall discuss this aspect explicitly in Chapter 2.

■ THE PIONEERS

The creation of ESRO can be attributed to two leading scientists: Edoardo Amaldi in Italy and Pierre Auger in France. Both men had been involved in the formation in 1954 of the European Center for Nuclear Research (CERN), which soon became a center of excellence in particle physics in Europe. CERN was created after the Second World War when the need for organizing big science at an international level became more and more necessary. After the launch of Sputnik-1 by the Soviets and the start of the American space program, it became apparent that space science was another field where the efforts of individual European nations were inadequate, even in the United Kingdom and in France where substantial domestic programs were already coming of age. Once the sharing of resources and expertise was accepted in principle, a strong political motivation was necessary to make it concrete.

Amaldi took the first step. He addressed a letter to several of his colleagues, including Auger, suggesting that an organization similar to CERN be set up for space science. Following an early discussion at COSPAR in Nice in January 1960, at which the British representative (Sir Harrie Massey) enthusiastically endorsed the concept, an informal meeting was held in Auger's Parisian apartment in February 1960, with representatives of France, Italy, Germany, Belgium, Holland, Sweden, Switzerland, and the United Kingdom participating. At several meetings later in 1960 the program of the future organization was discussed and an inventory of the launching means, test and integration facilities, and tracking stations of the potential Member States was established.

During that year the basis of ESRO was laid down and its main elements were specified: a headquarters (the Seat); an engineering center for integration, testing, and preparation for the launching of satellites and rockets (later called the European Space Technology Center, ESTEC); a network for satellite tracking and telemetry stations (ESTRACK); a center for data analysis (the European Space Data Center, ESDAC); a small research laboratory (the European Space Laboratory, ESLAB) to be set up near ESTEC; a launching base (ESRANGE). Also discussed were the main bodies of the future organization: a Council to define the policy and the administrative rules on which each Member State would be represented, and a Scientific Committee which would examine all proposals for research.

The decisions made by the representatives in these early meetings reflected an absolute desire to define ESRO as a purely scientific research organization, as free as possible from detailed governmental interference. This principle still governs the management methods applied to the scientific activities of ESA. The discussions at that early stage focused, for the most part, on scientific and technical issues. In particular, the delicate question of the launching means necessary to place the future satellites in orbit was carefully avoided as being too political: a European launcher would have the potential for military as well as scientific uses. Nevertheless, from active discussions of that issue in government circles emerged the concept of a European Launcher Development Organization (ELDO).

At the end of this early phase, a preparatory commission was set up and Auger was put in charge of its organization. At its first meeting on 13–14 March 1961 in Paris, a Bureau was elected which consisted of Sir Harrie Massey (U.K.) as president, Luigi Broglio (Italy) and Henk Van de Hulst (the Netherlands) as vice presidents, and Pierre Auger (France) as executive secretary. Simultaneously, two working groups were set up, one for scientific matters and the other for administrative and financial matters. In the following weeks the scientific program was formulated and published in the now famous ''Blue Book.''

In brief, a difference in attitude became apparent between the larger and the smaller countries. The larger countries (France, Germany, Italy, the United Kingdom) were most interested in projects of a size beyond what their own means permitted them to do; the smaller countries, with reduced or nonexistent means, were interested in setting up a sounding rocket and a small satellite program. For the smaller countries, the organization would serve the role of a space agency which they could use for their own needs, while some of the larger countries looked at it as a necessity but also as a competitor to their own national organizations. Hence, the program, as described in the "Blue Book," was built around a set of big and small satellites plus a sounding rocket program.

■ THE BUDGET

An important issue was the budget. The U.K. representatives, on the basis of their recent experience with CERN, took a very firm stance against an open-ended approach, a blank check, although they recognized the necessity of a budget of a scale sufficient to justify a European collaboration. The concept of capping the budget was eventually recognized by all as acceptable, and a ceiling of 306 million accounting units was fixed for the first eight years of existence of ESRO.[2] Provision was made, however, for the Council to adapt this ceiling to future needs by a unanimous decision of all Member States at the triennial determination of the level of resources. This measure was proposed—and strongly pushed by the United Kingdom—in order to protect national interests against a possible overrun of the budget of ESRO. The contributions of the Member States were based on their average net national income; however, no Member State was to contribute more than 25 percent of the total. The consequences of these financial agreements are still reflected today in the financial rules governing the mandatory activities of ESA, including the Science Program. Table 1 shows the first financial plan of ESRO and the evolution of ESA's budget is shown in Figure 1.

Table 1. The first estimate of the yearly ESRO budget established in 1961 (converted into millions of dollars)

Year	Payloads	Launch costs		Vehicles	Tracking and data handling	Fellowships	Total
		Rockets	Satellites				
1	8.1	2.8	0.0	0.3	0.6	0.3	12.1
2	14.0	2.0	0.0	0.8	7.0	0.3	24.1
3	18.8	1.6	0.0	0.8	8.1	0.3	29.6
4	17.9	1.6	3.6	4.4	2.8	0.3	30.6
5	17.6	1.6	5.3	6.1	3.0	0.3	33.9
6	17.6	1.6	11.1	12.0	3.0	0.3	45.6
7	17.6	1.6	11.2	12.0	3.0	0.3	45.7
8	17.6	1.6	11.2	12.0	3.0	0.3	45.7

Source: Sir Harrie Massey and M. O. Robbins, *History of British Space Science* (Cambridge: Cambridge University Press, 1966).

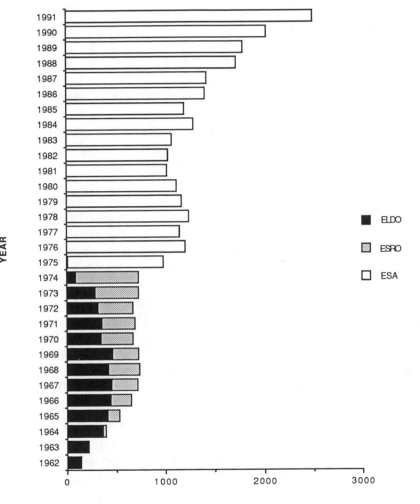

Figure 1. Evolution of the annual budgets of ESRO, ELDO, and ESA between 1962 and 1991. All figures are given in millions of accounting units (MAU) at the 1991 level. In 1993 one AU was equivalent to $1.30.

■ GEOGRAPHIC DISTRIBUTION

Considerations other than those of a purely scientific nature had to be taken into account early on. An issue that sparked heated debates among the Member States was the location of the various scientific, technical, and administrative establishments. This was the first occasion when the concept of geographic distribution was tacitly recognized as one of the key rules of the future organization. In hindsight, it can be seen that without this rule ESRO and ESA would have been more difficult, if not impossible, to create. Considerations of geographical distribution have played a significant role in the overall success of the enterprise.

The location of the headquarters was first proposed to be the same as that of the technical establishment, ESTEC, for the sake of coordination between the main scientific and technical activities and their administration. Even now, some consider this option preferable to having two separate locations.[3] The United Kingdom offered to host the combined ESTEC-Seat establishment in Bracknell near London, but this proposal failed as it was considered giving the United Kingdom too big an advantage. Eventually, after a vote of heads of delegations in April 1962, Paris was accepted, with ten in favor and two (Norway and Sweden) against, for the Seat alone, on the grounds of its proximity to the ELDO Headquarters, already located in the French capital.

The Delft Technical University (Holland) was chosen for ESTEC, with a majority of six votes against four in favor of Brussels. Later, in 1965, ESTEC was moved to Noordwijk, where more space was available, and which was closer to Schiphol Airport and within ten kilometers of the University of Leiden. ESTEC was first installed in temporary buildings, which were destroyed by a fire on 14 October 1966, and then in more solid ones in 1968.

Similarly, a majority of eight votes decided in favor of Darmstadt (West Germany), against four in favor of Geneva, for the location of ESDAC, which later became ESOC, the European Space Operation Center. It was also discussed whether to install ESLAB, the small research laboratory, within a European Space Laboratory for Advanced Research or a European Space Research Institute, proposed by Italy to be located in Frascati near Rome.[4] This proposal was not supported by the United Kingdom in particular, which favored locating ESLAB close to ESTEC. Italy (Broglio), however, insisted that a small laboratory of some kind be installed in Italy, a proposal which eventually was accepted. No agreement could be reached on what should exactly be the scope and the size of this establishment, a decision which was then left to the Council.

Sweden eventually hosted the sounding rocket launch facility (ESRANGE) in Kiruna.

■ THE APPOINTMENT OF SENIOR STAFF

Another issue requiring delicate political discussions was the filling of the key posts of the organization: Director General; Director of Administration; Scientific Director, responsible for ESRIN, ESLAB, and ESDAC; and Technical Director, responsible for ESTEC. Concerning these appointments the Member States showed a determination comparable to their interest in the location of the establishments.

The post of Director General was filled without difficulty: Pierre Auger was elected unanimously. For the other posts, the Heads of delegations acted as an appointments commission. The post of Technical Director was filled also relatively easily by Freddy Lines from the United Kingdom, as a consequence of his indisputable technical capability acquired at the Royal Aircraft Establishment. Reimar Lüst was appointed Scientific Director on a part-time basis, continuing his association with the Max Planck Institute for Extraterrestrial Physics in Garching.[5] Reporting to him were Hermann Jordan from West Germany as Head of ESRIN, Ernst Trendelenburg, also from West Germany, as Head of ESLAB, and Stig Comet, from Sweden, as Head of ESDAC. The appointment of the Director of Administration was more difficult. The United Kingdom pushed the candidacy of Thomas Crowley, against the wishes of Auger, who had hoped that his assistant in the preparatory commission, Jean Mussard, would get the post. Eventually, Mussard was appointed to the newly created position of Secretary of the Council, and Crowley was appointed Director of Administration on the grounds that the United Kingdom was the strongest financial contributor and had no establishment on its territory. As we can see, political and nationalistic considerations have played a major role in the setting up of the main bodies of ESRO.

■ THE ESRO CONVENTION

The ESRO Convention and the financial protocol were signed on 14 June 1962 in Paris by all the original members of the Preparatory Commission except

Norway (Belgium, Denmark, France, Italy, the Netherlands, Norway, Sweden, Switzerland, United Kingdom, West Germany), with Austria in addition. The Convention instituted the Organization under a Director General and a Council to which each Member State would nominate two representatives. The Council should elect its chairman and two vice chairmen, and should meet at least twice a year. The Convention would only come into force when signed by at least six Member States representing 75 percent of the contributions to the total budget and including the countries in which the future ESRO establishments would be installed.

By 13 March 1964 all countries except Italy had ratified the Convention, which entered into force on 20 March. A meeting of the Preparatory Commission was convened at which it was agreed that, in order not to waste more time, Italian delegates could participate in the meetings of the Council with no voting power until Italy had ratified the Convention. In the meantime, Italy could participate in the scientific activities and Italian nationals could be eligible for staff positions, while no decisions about the European Space Research Institute (ESRIN) would be possible without Italian agreement. The first meeting of the Council took place on 23–24 March 1964. Observer status was given to Austria and offered to Norway. The first chairman was Sir Harrie Massey, unanimously elected after a proposal from the Dutch delegation. Alexander Hocker (West Germany) and Henk Van de Hulst (the Netherlands) were elected as vice chairmen. At that meeting, the Council created its Scientific and Technical Committee and its Administrative Committee.[6]

The role of the Council is to look after the overall policy and programs of the Organization, while the daily management is left to the Executive. However, before ESA came into existence, the overall European space policy was established at the level of what was then called the European Space Conference (ESC), which coordinated the policies of ESRO, ELDO, and the European Conference for Space Telecommunications. The ESC was convened nine times between 1966 and 1977. When ESA was created the ESC ended its activities. Its role was then filled by the ESA Council at the ministerial level.

Council meetings at ministerial level offer the opportunity to redefine the major goals of the Organization. They constitute the forum where main programs are initiated and their future evolution is debated. Noteworthy was the November 1968 meeting of the ESC in Bad Godesberg, when ministers agreed to the principle of merging ELDO and ESRO and forming a single European Space Agency with a single space program. This meeting was the first to discuss the concept of optional programs, which were created in order to allow the Member States who wished to participate in the development of a European launcher and of application programs to do so. The complementary concept of a minimum basic program, mandatory for all Member States, also dates back to that meeting. Giampietro Puppi, then chairman of the ESRO Council, was given the task of turning these ideas into more concrete proposals with associated budgets.

Also very important was the ESC meeting in December 1971 in Brussels, when all the elements of the future program of ESA were agreed upon in what is referred to as the "first package deal," and when the main principles of the future ESA Convention were laid down. These agreements of principle were transformed into a detailed plan at the following ESC meeting in December 1972 and in July 1973, when the "second package deal" was approved.

Later, following the successful implementation of this package deal, the future activities of ESA were discussed at the first two ministerial Council meetings, in 1985 in Rome and in 1987 in The Hague. At these meetings the ministers accepted in principle the European participation in the space station, the Columbus program, the development of the new version of the Ariane launcher, Ariane-5, and the initiation of the Hermes European space plane. Last but by no means least, at the Rome meeting the Horizon 2000 program for space science was unanimously endorsed as the core of the mandatory scientific activities within ESA, with the commitment of principle to increase the science budget by more than 50 percent over ten years. We shall come back to these aspects in the following chapters. The ministerial Council meetings held in Munich and Granada in 1991 and 1992 respectively, coming after the collapse of the USSR and coinciding with the end of the cold war, have been other turning points in the ESA's activities.

■ THE EUROPEAN LAUNCHER DEVELOPMENT ORGANIZATION

Simultaneously with the setting up of ESRO, the two main European space powers, the United Kingdom and France, were seriously considering developing the launching capability necessary to place the future European satellites into orbit. Without their own, ESRO had to rely essentially on American launchers. The United Kingdom had developed the Blue Streak, an element of its future ballistic missile, but the project was abandoned in April 1960 before it could be used; it was already obsolete in comparison with the more modern American Polaris and Minuteman missiles. The U.K. government, which had spent nearly £100 million on the engine, thought that the Blue

Streak might constitute an excellent first stage for a future civilian launcher. Similarly, the French were developing their future nuclear missile, together with a series of powerful sounding rockets, and thought that their experience in this field could be exploited in the development of a satellite launcher.

While ESRO was created by scientists in order to meet scientific needs, ELDO was created by governments interested in taking advantage of their technological expertise and of their earlier financial investments, either in the military or in the civilian sector. There seems to have been little contact between the two groups discussing cooperation in space science and in the development of a launcher. The scientific community was reluctant to be involved in the launcher activities, probably because of the potential military implications and the fear that space science budgets might be swallowed up. As a consequence, the negotiations for the establishment of the two European space organizations proceeded nearly independently.

Following a confidential talk with the British Prime Minister, Harold Macmillan, at Rambouillet on 28 January 1961, the French President, Charles de Gaulle, realized that Europe could become the third space power. On 30 January 1961, at a meeting at the Maison de l'Europe in Strasbourg chaired by Sir Peter Thorneycroft, the U.K. Minister of Aviation, the basis of the future European Launcher Development Organization (ELDO) was established. It would be responsible for developing a launcher capable of placing about one ton into circular orbit at five hundred kilometers. But it took longer before the project could be finalized. The Convention was signed on 30 April 1962 among seven participating states: Australia, Belgium, France, Germany, Italy, the Netherlands, and the United Kingdom. It came into force on 29 February 1964.

The launcher, called Europa, had three stages: the first was the U.K. Blue Streak; the second, Coralie, was developed by France; and the third was the responsibility of a consortium of European countries led by Germany. In addition, Italy was made responsible for the development and construction of the first series of satellites, Belgium for the equipment of the ground stations necessary for intermediate guidance, and the Netherlands for the long distance telemetry system and the associated ground equipment. The launching base was located at Woomera in Australia.[7] Remarkably absent from the work of the preparatory group were Sweden and Switzerland, two neutral countries which were reluctant to associate themselves with an enterprise that relied too heavily on military developments, although it was well understood that the activities of ELDO would be exclusively of a peaceful nature. Eventually Switzerland, together with Denmark, which was involved in the preparatory negotiations, obtained the status of observer. The budget of the organization added the contributions of each participating country in support of their respective tasks, quite a significant deviation from the GNP-based financial system of ESRO.[8] It was voted every year with a two-thirds majority, and a positive vote was required from those Member States contributing 85 percent.

Like ESRO, ELDO was governed by a Council. In addition, a Secretary General, a Technical Director, and an Administrative Director constituted the main executive posts, complemented by a minimum of staff. Its headquarters were in Paris. The task of the Secretary General was fundamentally different from that of ESRO's Director General. He had only the vague power of coordinating activities which, for the most part, were conducted by already existing and powerful national organizations, in particular in France and in the United Kingdom. The responsibility for industrial contracts was assigned to the participating governments and not to the Executive. The ELDO council met for the first time on 5 May 1964. The Italian Ambassador, Renzo Carrobio di Carrobio, who had been acting as Secretary General of the preparatory group, was unanimously designated as Secretary General.

Unfortunately, very soon after it was born, ELDO evidenced grave problems which arose from the very principles which governed it. The organization was deprived of any genuine technical and management responsibilities, which were distributed to the various agencies of its Member States. Each Member State was responsible for its own part. There was no Prime Contractor in charge of the overall management of the program. Although the first stage of the launcher was successfully tested—in fact the Blue Streak never failed in the whole series of test flights—the system tests of the first- and second-stage assembly never worked. Europa never was able to reach orbit or to launch a satellite: it failed completely in its mission. In addition, the first financial estimates, established in 1962 at a level of £70 million over a period of five years, very soon proved to be largely insufficient. At the same time, the schedule was continuously slipping. Furthermore, the purpose of the launcher had been poorly defined. It was in this context that the European Space Conference decided in 1966 to increase Europa's capability so that it could place about 200 kg into geostationary orbit, a capacity which better fitted future European needs, in particular in the area of telecommunication satellites. The requirements of the development of this new version nearly tripled the original budget, and the budget had to be increased again in 1969 as a consequence of technical difficulties and subsequent delays.

The new launcher, Europa-2, was tested on 5 November 1971 from Kourou. This was its first and unfortunately last flight. The cause of the problem was traced to an electrical interference in the German third-stage system. The launcher's spectacular failure forced the Member States to review the management procedures of the program in relation to the respective responsibilities of the ELDO Executive, the Member States, and industry. Some Member States were already questioning whether it was really necessary to pursue the development of a European launcher. On 1 May 1973 the Council of ELDO decided to stop the development of Europa-2. A new version, Europa-3, was under discussion, but it never flew above the drawing board where it had been under study since 1968. Nevertheless, its concept laid the basis for the development of Ariane (Figure 2), at that time called L3-S (3-Stage Launcher of Substitution), whose construction was part of the second ESA package deal already mentioned. Deprived of its main *raison d'être,* ELDO ceased its activities in 1974 when all its personnel were transferred to ESRO. In 1975 ESA came into existence, with as its main task the development of Ariane under the financial and technical responsibility of a single organization, CNES from France. This arrangement was intended to avoid repetition of the fundamental errors which had marred one of the most difficult experiences of the early years of European cooperation in space.

Why did ELDO fail in comparison with ESRO? ESRO provided facilities and services to the European scientific community. Its program was kept as much as possible free of political interference, being defined, reviewed, and used by the scientists themselves, but being managed through the Agency. On the contrary, ELDO, placed directly under the auspices of its Member States, was deprived of any central management capability. The example of ELDO must be remembered in the future when setting up other research and development organizations of multinational character.

■ THE ESRO CRISIS AND THE BIRTH OF ESA

In 1965 the ESRO Council approved an eight-year program which included three hundred sounding rockets, two series of small satellites in low earth orbit, and two large satellites. The first four years of ESRO's existence were devoted to setting up the main installations and establishments and in recruiting the staff. In 1966 the three main establishments were already in operation. The launching base, ESRANGE, was built at Kiruna in northern Sweden, and the network of tracking stations was progressively entering into service. At the end of 1968 ESRO had succeeded in launching one hundred sounding rockets and, with the help of American launchers, its first three small satellites : ESRO-2B, devoted to the study of cosmic rays and solar X-rays, on 17 May 1968;

Figure 2. Ariane on its launch pad at Kourou. The launcher has been designed specifically to place satellites in a geostationary transfer orbit.

ESRO-1A, devoted to the study of auroras and of the ionosphere, on 3 October 1968; and HEOS-1, devoted to magnetospheric research and the study of Sun-Earth relations, on 5 December 1968. The list of spacecraft at the end of the book gives a brief description of the main characteristics of these missions.

In spite of these early successes, in this same period ESRO underwent its first major crisis. The first cause of that crisis was ESRO's inability to establish proper technical and financial definitions of the first two large satellites of its program.

The first was a Large Astronomical Satellite (LAS), for the observation of celestial sources in the ultraviolet between 3500Å and 912Å. The spacecraft was to be stabilized along three axes to within one arc second, and the diameter of the telescope was in the range 50–80 cm. By 1965 contacts between ESRO and ELDO had improved considerably and it was envisaged that the satellite could be launched by Europa from the new launching range in French Guyana, which allowed a substantial gain in weight.

The second large project was even more ambitious. It was envisaged as either a lunar satellite, a station on the Moon, a planetary mission of some sort, or a comet flyby. Eventually, the comet project was recommended by an ad hoc group of scientists chaired by Pol Swings from Belgium, one member of which was Ludwig Biermann from Germany, a famous name in comet science. Another suggestion was made for an international flyby mission to Jupiter. But none of these missions ever came to fruition.

The high degree of ambition of these first two large projects illustrates the immaturity of the young organization. The crisis was latent. At the same time, the Member States were struggling in the Council to keep the budget within acceptable limits. The inability of the Council to agree on the triennial level of resources forced ESRO to operate on yearly and sometimes monthly budgets. There was growing dissatisfaction in the Member States as well as in scientific circles, where ESRO was seen as a high-cost organization, giving poor value for money, with project costs escalating and schedules continuously slipping. Even the two moderate-sized satellites, TD-1 and TD-2, could not be produced within the accepted budget limits. In 1967 Spain requested that its contribution be reduced below the proportion of its normal GNP share.

Underlying the crisis was also the increasingly heard criticism that ESRO was too much of a scientific preserve, while its expensive technical and engineering facilities could be of great value for other potential cooperative European programs in telecommunications and other space applications. The pressure was growing to amalgamate ESRO and ELDO into a single body in charge of implementing a single coordinated space policy, incorporating science as well as application programs.

Following the recommendations of a group of experts chaired by Jan Bannier from the Netherlands, the European Space Conference, at its second meeting in Rome in July 1967, set up an Advisory Committee to review the future of the organization and its programs, in particular the two large projects, the LAS and the comet mission. The committee, chaired by Jean-Pierre Causse of France, issued its report in January 1968. It was discussed at the Bad Godesberg Space Conference the following November. The report called for the fusion of ESRO and ELDO, and for the first time the name European Space Agency was suggested for the combined organization. The LAS, which had requested more than 40 percent of the available resources of ESRO for 1968–1972, was abandoned, and so was the comet mission.

Two projects of comparable objectives, however, would be carried out later and in a different framework. First, the LAS was redefined and undertaken as a joint program between NASA, the United Kingdom, and ESA, and was renamed the International Ultraviolet Explorer, IUE (Figure 3). It was launched in 1978 and has been operated since then jointly by NASA and ESA. It is one of the great successes of space astronomy today. Second, a flyby mission to Halley's Comet, Giotto (Figure 4), was decided on by ESA's Science Program Committee in July 1980. It has been one of the most spectacular European successes in the exploration of the Solar System and in space science.

The Causse report recommended that the new agency, in addition to scientific satellites, should carry forward a program of application satellites and should develop launch vehicles, together with a substantial technology research program in support of such activities. Another senior committee, chaired by Giampietro Puppi, also chairman of the Council at that time, was established to examine the recommendations in more depth. In the meantime, France and Denmark denounced the ESRO Convention in order to put pressure on the other nations to enlarge the mission of the organization toward applications. For a few months ESRO's future was on very uncertain ground.

Puppi's proposals, the essence of the "first package deal," were discussed and accepted at the European Space Conference in December 1971 in Brussels. Their main effects on ESRO's science activities were to phase out the sounding rocket program by mid-1972; to end the current activities of ESRIN in 1973; to decrease the level of the science budget to no less than 27 million AU— Puppi had proposed a level of 35 million AU, which was subsequently reduced by the Council; to transfer the responsibilities of ESRANGE to Sweden in

Figure 3. The International Ultraviolet Explorer (IUE) launched in 1978 can be considered a reduced-scope version of the Large Astronomical Satellite, the first large project of ESRO. On 10 March 1993 ESA celebrated the fifteenth anniversary of the spacecraft in orbit.

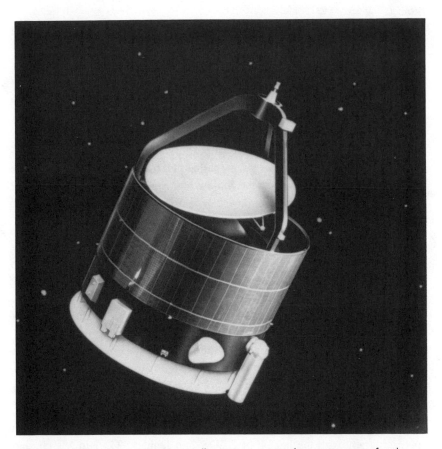

Figure 4. ESA's Giotto mission to Halley's Comet. Nearly twenty years after the elimination of the second large project from the ESRO program, Giotto became the—modest—realization of ESRO's original ambition. It was successfully launched by an Ariane-1 rocket on 2 July 1985 and performed its mission flawlessly. This was the first time that ESA launched a scientific mission with its own means, and it was ESA's first interplanetary mission. It was also the first noncommercial mission of Ariane. Giotto survived its encounter with Halley on 13–14 March 1986 and was redirected on a new orbit in July 1990 using, for the first time in history, the Earth for a gravity-assisted maneuver. It encountered Comet Grigg-Skjellerup on 10 July 1992 at a record minimum distance of 200 km. Giotto is the only spacecraft to have encountered two comets. It will return to the vicinity of the Earth in 1999.

July 1972, while maintaining a reduced sounding rocket program as a special project.

The Council clearly indicated its willingness to favor the introduction of application programs at the expense of scientific activities. France and Denmark immediately withdrew their denunciation. The first package deal was then approved. Such a reorientation required a profound revision of the Convention in order to allow the Member States to participate in the application programs according to their wishes and not within the budgetary constraint of the GNP rule in effect at ESRO.

No fewer than three subsequent European Space Conferences were necessary in order to formalize this agreement and place ESA on its firm track. The first, in December 1972, defined the content of the program, called the "second package deal," which can be summarized as follows:

- A European Space Agency would be established which would combine the functions of ESRO and ELDO and whose role would be to coordinate and progressively integrate into a single European space program the various individual national space activities.
- The French rocket design, then called L3-S, later renamed Ariane, was adopted as the future European launcher in preference to Europa, developed by ELDO.
- The Europeans would cooperate with NASA in the development of their Space Transportation System, the Space Shuttle, by developing Spacelab.
- A maritime navigation satellite would be developed as part of the applications program.

In retrospect, 1972 can be seen as a very successful year for the European space program, with the launch of HEOS-2 in January, TD-1 in March, and ESRO-4 in November. The second conference in July 1973, thanks to its chairman, Belgian Minister Charles Hanin, was a success: a general agreement was obtained on all points of the package deal. A few weeks later in Washington, on 24 September, the agreement on Spacelab was solemnly signed. In the following months the other agreements were formalized, in particular that concerning Ariane. The third conference, held in Brussels on 15 April 1975, approved the terms of the Convention, which was signed in May by ten Member States.[9]

After ten years, Europe had finally acquired the framework it had been looking for in order to coordinate, and possibly later integrate, all its common and national space programs. Roy Gibson became the first Director General of the new Agency. The Convention was slow to be ratified, however. This

Table 2. ESRO and ESA scientific satellites launched as of 1993

Spacecraft	Mission	Launch date	End of life	Launcher
ESRO-2B	Cosmic rays and solar X-rays	17 May 68	09 May 71	Scout
ESRO-1A	Polar ionosphere and auroral phenomena	03 Oct. 68	26 June 70	Scout
HEOS-1	Solar wind and interplanetary medium	05 Dec. 68	25 Oct. 75	Delta
ESRO-1B	Polar ionosphere and auroral phenomena	01 Oct. 69	23 Nov. 69	Scout
HEOS-2	Polar magnetosphere and interplanetary medium	31 Jan. 72	02 Aug. 74	Delta
TD-1	Ultraviolet astronomy	12 March 72	04 May 74	Delta
ESRO-4	Ionosphere and solar particles	22 Nov. 72	15 April 74	Scout
COS-B	Gamma-ray astronomy	09 Aug. 75	25 April 82	Delta
GEOS-1	Magnetosphere	20 April 77	23 June 78	Delta
ISEE-2*	Magnetosphere and Sun-Earth relations	22 Oct. 77	26 Sept. 87	Delta
IUE*	Ultraviolet astronomy	26 Jan. 78	?	Delta
GEOS-2	Magnetosphere	14 July 78	25 Aug. 85	Delta
Exosat	Cosmic X-rays	26 May 83	09 April 86	Delta
Giotto	Flyby of Halley's Comet on 13 March 1986 and of Comet Grigg-Skellerup on 10 July 1992	02 July 85	?	Ariane
Hipparcos	Astrometry, positions, and proper motions of stars	08 Aug. 89	15 Aug. 93	Ariane
Hubble Space Telescope*	Long-term optical and ultraviolet observatory in space	24 April 90	?	Shuttle
Ulysses*	Study of heliospheric environment as a function of solar latitude	06 Oct. 90	?	Shuttle

*ESA-NASA project.

Note: For more information on these missions, see the list of ESRO and ESA spacecraft, page 147.

was done on 30 October 1980, when all the Member States had cleared their participation with their respective Parliaments.

This short overview illustrates the difficult beginnings of what can be considered as the first multinational space organization. ESRO was a purely scientific organization and its Convention did not offer the possibility of participating in application programs. The failure of ELDO and the need to start application programs, whose dimensions and ambitions were far above the limited financial capabilities of ESRO, led the Member States to redefine and enlarge its scope. ESRO could not resist the pressure, and had to lead the way to ESA. After several years of discussions, learning, and organizational work, success eventually came through. This success can be measured through the number of scientific and application satellites so far launched and operated (Tables 2 and 3). Another success of ESRO and ESA has been in fostering a European spirit, strengthening the willingness of the scientists and their Member States to pool their resources and engage in a common effort in space science and in technology.

ESRO itself succeeded in launching seven medium-sized satellites in the period 1968–1972 and in defining the more ambitious missions that were to form the present successes of ESA. It was forced, however, to restrain its

Table 3.   ESA applications satellites launched as of 1993

Satellite	Launch date	Mission	Launcher
Meteosat 1	23 Nov. 1977	Meteorology	Delta
OTS-2	11 May 1978	Communications	Delta
Meteosat 2	19 June 1981	Meteorology	Ariane
Marecs-A	20 Dec. 1981	Maritime communications (Pacific)	Ariane
ECS-1	16 June 1983	Communications	Ariane
ECS-2	04 Aug. 1984	Communications	Ariane
Marecs B-2	10 Nov. 1984	Maritime communications (Atlantic)	Ariane
ECS-4	15 Sept. 1987	Communications	Ariane
Meteosat 3	15 June 1988	Meteorology	Ariane
ECS-5	21 July 1988	Communications	Ariane
Meteosat 4	07 March 1989	Meteorology	Ariane
Olympus-1	12 July 1989	Communications	Ariane
Meteosat 5	02 March 1991	Meteorology	Ariane
ERS-1	17 July 1991	Global ocean/ice, regional land observation	Ariane
Meteosat 6	19 Nov. 1993	Meteorology	Ariane

original ambitions. Its first two large projects, the LAS and the comet mission, were abandoned because their designs were unrealistic at that time and because of ESRO's inexperience in management and budget estimation. Within the limit of a capped budget, nearly one order of magnitude smaller than its American equivalent, ESRO and then ESA had to abandon, right at the outset, the race to planetary exploration, which was clearly beyond its financial means. Giotto, a somewhat modestly sized mission, launched in 1985, was the first, and very successful, attempt by Europe to enter into deep space by its own means.

Between an obvious lack of experience and a strong desire for independence, ESRO had a very narrow path to follow. It was totally dependent upon the United States for the launch of its satellites. The interest of the Member States in taking maximum advantage of the technological advances permitted by the space program led them to envisage a progressively expanding set of missions. This interest and the desire for self-sufficiency can be identified as major reasons for their continuing support of ESA. Political support was essential through the years if that common effort was to proceed and to progress. In that respect, the role of the European Space Conference and of the Council meetings at ministerial level should not be underestimated. They helped in resetting the overall policy and in defining the new directions in which ESA would proceed. They have proved to be an essential element in the successful running of the Agency.

2

GOVERNING PRINCIPLES

An organization like ESA can work only within a set of precise rules. The Convention defines the legal framework for the definition of the European space policy and for the management of the Agency and its projects. Within this framework, the Executive retains a large amount of initiative, as do the Member States through the Council and the various committees. Science policy and industrial policy have played a very important role in the development of a European spirit and in the process of European integration. The Member States do control the implementation of that policy, and the relationships between them and the Agency are naturally a delicate issue. ESA is probably one of the most open space organizations, one in which the users and the Member States are continuously informed, one they can check carefully, if they wish, to make sure their desires are properly implemented, the missions properly managed, and their financial contributions properly used.

The ground rules of ESA are very clear: all Member States shall participate in the mandatory activities and shall contribute to the fixed common costs (Article I of the Convention). All authority flows from the Council and the Director General. The Council is the sovereign legislative body, composed of delegates of the Member States; the Director General is the executive officer and is nominated by the delegates with a two-thirds majority—although the normal procedure is to elect him unanimously or by acclamation of the Council. The Agency enjoys immunities and privileges customary to international organizations. The Council meets as required either at delegate or ministerial level. The Council of ESA took up its duties in mid-1975.

Article II of the Convention states the tasks of ESA: ''The purpose of the Agency shall be to provide for and to promote for exclusively peaceful purposes, cooperation among European States in space research and technology and their space applications with a view to their being used for scientific purposes and for operational space applications systems.'' Thus the two main branches of the Agency's activities are space research and technology and space application systems. The first branch corresponds to the former ESRO mandate while the second is peculiar to ESA.

Article V of the Convention, which deals with activities and programs, specifies that the scientific and technological research work belongs to the mandatory activities of the Agency, in which all Member States participate, while the space application systems belong to the optional activities, in which in principle all Member States participate except those which formally declare themselves not interested in participating.

▪ THE MANDATORY SCIENTIFIC PROGRAM

The Scientific Program is the core of the Agency's mandatory activities, which also include technology research and general administration. This is the program for which ESRO was established. To underscore its privileged position, the Science Program Committee (SPC), to which the Council shall refer any matter relating to the mandatory program, is the only program committee under the authority of the Council to be specifically identified in the ESA Convention.

One characteristic of the way ESA manages the Scientific Program, and one way in which ESA differs from other agencies like NASA, is that the payload elements selected by the SPC—in competition—to fly on an ESA spacecraft are financed by the nations themselves rather than from the ESA science budget, which funds only common facilities. This has been the ESRO and ESA policy throughout the years and has encouraged a high degree of involvement and initiative within the scientific groups, which have not been able to take funding for granted. As far as the data are concerned, it is the task of ESA to ensure that all scientific results be published after prior use by the investigators responsible for the payload elements.

The SPC

The SPC is composed of delegates of the Member States with specific competence in scientific matters. The general procedure is for each delegation to the SPC to be composed of two representatives, one of whom should be a scientist. This has worked very well over the years, the two representatives sharing the floor depending upon whether the discussion is of a scientific or a political or administrative nature. Sometimes amusing situations develop, as when Giuseppe Occhialini, who was sitting in the Italian delegation, pulled down the arm of the politician of that same delegation who was going to vote yes on the matter being discussed!

The SPC has total authority over the scientific projects: preliminary study, approval, financing, and monitoring of the development of the common facilities and of the operational phases of a mission. However, the level of its budget is set by the Council, which has the authority to determine the multi-year level of resources and to adopt the annual budget. In fact, the Council has been very cooperative with the scientists, even within difficult constraints, and has often gone obligingly along with what one might call a "creative budgeting" approach, at times when the SPC tried to find ways out of delicate financial impasses. This was possible because of the complete openness of the process, a quality well understood by both the SPC and by the Executive to be essential to obtain an agreement by the highest authority: the Council. The rules of procedure of the SPC are typical of any such body. Two issues should be highlighted, however, because of their particular relevance to the ESA Scientific Program.

The first concerns the procedure for deciding on projects. The Convention stipulates that while a scientific project shall be approved by a majority of all the Member States, it requires a two-thirds majority to change these decisions. This proviso is unique to the Scientific Program and does not apply to the optional programs. Its effect is to add credibility and solidity to a project, protecting it from unwarranted changes which might cause time delays and cost increases. It also puts heavy pressure on the scientific and industrial teams involved in the studies prior to approval to come up with a clear design, clean interfaces, and credible schedules and cost estimates. The result has been encouraging: all scientific projects have in general contained increases well within 25 percent of the approved costs established at the end of Phase B, which is a significant achievement in high technology (Table 4). Exceptions are the Hubble Space Telescope, whose cost to ESA has more than doubled with respect to the original estimate as a consequence of interface changes imposed by NASA and of launch delays, and the Infrared Space Observatory, which has suffered technical or scientific difficulties. Higher levels of modification have also sometimes been necessary because of totally unpredictable events, outside the control of the teams involved in the development of the missions—for instance the change of the launcher in the case of Exosat and Giotto—but such modifications must be well documented and justified to achieve the two-thirds majority required for approval.

The second issue refers to the relationship between the ESA Science Program and the national programs. The Convention stipulates (Article II) that

Table 4. Cost evolution (in MAU) of three science reference projects

Project	Year of approval	Year of launch At approval	Year of launch Actual	Economic condition	Cost Phase A	Cost Phase B	Cost % A to B	Cost At completion	Cost % B to completion
Ulysses[a]	1977	1983	1990	1979	N.A.[b]	103.2	NA	118.6	15[c]
Giotto	1980	1985	1985	1981	108.1	129.7[d]	20	141.8	9
Hipparcos	1980	1987	1989	1982	193.5	243.8[e]	26[f]	296.0[g]	21

a. ESA-NASA cooperative mission.

b. Mission redefined at end of Phase A when NASA dropped its spacecraft in 1981.

c. Launch delayed by 4.5 years after the Challenger accident.

d. Phase A was a "crash action" based on a satellite design no longer available. The payload was increased relative to Phase A and improved comet modeling led to a better technical definition.

e. Phase B extended because of fundamental technological problems.

f. The payload, usually paid by the scientific institutes, was introduced as an ESA responsibility in the course of Phase B.

g. Launch delayed due to problems with the Ariane launcher, and more funds were necessary to operate the satellite, whose orbit was non-nominal owing to the failure of the apogee motor.

the programs should be coordinated, and that the national programs should be integrated as completely as possible into a European program. That responsibility falls naturally on the SPC, by delegation of the Council.

In fact, effective coordination cannot be imposed but must evolve naturally. Also, a European program is not merely the sum of all haphazardly approved projects, but, if coordination is to be exerted, must be preplanned and made public. Coordination, therefore, became effective only with the emergence of the long-term program in space science, Horizon 2000, and with the periodic organization of special meetings of the SPC exclusively dedicated to the presentation of national programs. These are known as the Capri meetings, because they usually take place on Tiberius Island where the SPC Italian delegate, Dr. Saverio Valente, has several times been elected mayor of the beautiful city which overlooks the Bay of Naples.

Financing

The Scientific Program budget is based upon mandatory contributions of the Member States. These contributions follow a key related to their respective GNPs. The two concepts, mandatory and GNP-related contributions, are central to the management practice governing the Science Program.

In fact, because of the role of the Science Program at the heart of ESA's activity—some call it the backbone or the brain of the organization—and as a source of technological innovation, each Member State is required to participate in it. It is the only program of the Agency, and the only one in Europe, with the permanent participation of all the Member States. Furthermore, as it is a science-oriented program, decisions therein have to be in the first place driven by science; political and industrial considerations should intervene only at a second stage. For these reasons, unlike the case of the optional programs, contributions are mandatory and their amount uniquely determined by a key reflecting the overall economic performance of each country and not a national choice of priorities. Through these two rules the Scientific Program is by and large, under normal conditions, shielded from unwarranted national pressures, as no individual nation can use the leverage of its financial contribution or of its membership in the program to further its own preferences. These two safeguards, which one cannot emphasize enough, are missing in the optional programs, where participation is voluntary and where contributions are tailored to each Member State's interests.

From the procedural point of view, the financing of the Science Program is determined through a five-year level of resources which the Council must approve unanimously and which is reviewed every three years. The annual budgets, which have to be consistent with the overall level of resources, are approved every year by the Council with a two-thirds majority.

The two conditions set out above, which have been strenuously defended throughout the years, have sometimes led to seemingly contradictory consequences. On the positive side, the choice of projects and of the payload complement is essentially based on arguments of scientific excellence alone and does indeed represent the preference of the majority of the Member States. Equally positive is the security and guarantee of financing, over each five years at least, which allows for effective forward planning. In fact, this stability is quite unique among the space science programs in the world. On the negative side, and in apparent contradiction to the above, one can mention the inflexibility of the financing. Indeed, the combination of mandatory and GNP-related contributions leads quite naturally to the fact that any change of the level of financing, either downward or upward, is difficult to bring about unless there is unanimous agreement. Hence, although funding is secure, increases are rare. Rare, but not impossible if one has an extremely convincing set of arguments. This was the case with the Horizon 2000 program, ESA's first long-term plan in space science.

▪ OPTIONAL PROGRAMS

As explained earlier, the strict rules of the mandatory activities were not amenable to the inclusion in ESRO of the large application programs which were essentially the product of national political and economic priorities. While in the Scientific Program science was itself the main force—albeit with an eye to the potential technological spin-off—and scientific excellence, rather than economic benefit or national prestige, was the prevailing parameter, it cannot be denied that the latter considerations were more important in the applications area. Therefore, it had to be accepted that each nation should be able to choose the level of its financial involvement according to its economic and industrial conditions. Hence the concept of ''optionality'' was introduced in the ESA Convention and through the two package deals of 1971 and 1973 a *menu à la carte* was established to which the Member States were free to adhere or not, without limits to the percentage of their contributions. The menu of optional programs included telecommunications, meteorology, and navigation satellites as well as Spacelab and Ariane.

The Convention promotes and encourages participation in these optional programs, and the Member States are automatically participants unless they specifically declare their wish not to participate. Also, the Convention stipulates that no proposal of any single nation can become an ESA program; only those which fulfill ESA's basic rules and aims can be accepted. Thus the carrying out of an optional program within the framework of ESA requires an acceptance by the majority of all Member States.

As regards financial contributions to optional programs, the Convention encourages a scale of contributions based on the GNP of each participating state, but it leaves open the possibility that the states may agree on a different scale. The Convention also stipulates that no participating state shall be entitled to withdraw from that program unless the cumulative cost overrun is greater than 20 percent of the initial financial envelope. This rule is now under debate in the Council following the ministerial meeting in Granada of November 1992.

The Convention, throughout all its articles, encourages the execution of the programs in accordance with its general provisions and the rules and procedures in force in the Agency, unless otherwise stipulated and accepted by the Council. There is an insistence on maintaining as much as possible a commonality in the management of both the mandatory and the optional programs, on maintaining the maximum of cohesion among the different programs, and on preventing a fragmentation which would be detrimental to the Agency itself and to the coherence of the European space effort.

Nowadays, the optional programs constitute more than 80 percent of the Agency's activities, evidence of how vital it was that the Convention of ESRO be modified to make way for these programs. Figure 5 shows the idealized flow chart for the planning and implementation of ESA's optional programs and demonstrates that a considerable amount of coordination and planning is required.

Optional programs need not deal only with application satellites or with launcher and infrastructure developments; they can also be scientific, although this possibility is generally not favored by the Member States because it contradicts the basic tenets of the Science Program. However, in some cases the "optionality" can be instrumental in securing additional resources—without modifying the basic level approved unanimously—and in carrying out additional activities without detriment to the main program. A typical case is the extension of the operations of an orbiting spacecraft beyond the nominal lifetime included in the budget. This has occurred for both the COS-B and GEOS missions. Similar extension of the IUE has been proposed on several occasions but rejected by the Member States. Another case would be the assign-

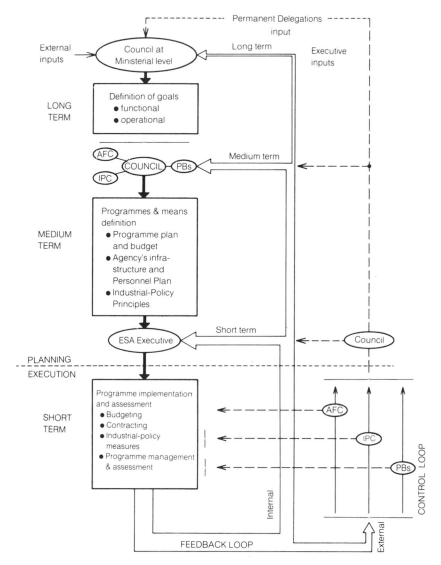

Figure 5. Flow chart for the planning and implementation of ESA's optional programs.

ment of a new mission to a spacecraft which has already accomplished its primary mission. Thus, an optional program was proposed to direct Giotto—although some of its instruments were damaged during the encounter with Halley's Comet—to intercept another comet, Grigg-Skjellerup, on 10 July 1992. This and the IUE proposal were rejected by the SPC, not because they were institutionally unacceptable but because in both cases it was argued that the Mandatory Program should stand by its own projects and cope with its own success.

▪ SCIENCE POLICY AND LONG-TERM GOALS

ESA's science policy aims at providing the scientific community in Europe with the best possible set of missions, as judged by the uniqueness and excellence of their scientific objectives. The fundamental rule of ESRO, and then of ESA, has been that the Agency exists to serve the scientists and that the science policy of the Agency has to be driven by the scientific community and not vice-versa. This principle has profound implications for the relations between the scientific community and the Agency. Throughout its history, the science policy has been defined by the scientists themselves and has reflected their aspirations. Though not every single scientist would recognize in ESA's Scientific Program his or her preferred project, there is a general consensus among European scientists that, by and large, the program has served them well, as it has served space science in Europe and science more generally.

Another concept related to this science-centeredness is that scientists should fund their own payloads under national auspices, while ESA will fund the procurement of the common services, usually the carrier, the launcher, and the operations. Thus was averted the danger that payloads would be selected on the basis of political pressures or assembled according to some administratively inspired scale such as that of just return which, as described below, governs ESA's industrial policy.

These ground rules had far-reaching consequences. Between the community and ESA was established a vital link, a sort of complicity, which was at the root of ESA's success in its projects and its policy. Scientists worked with ESA in a fully committed way, with no guarantee of compensation or personal profit, for the achievement of the best possible program for Europe. Groups cooperated to optimize their scientific effort, and were able to convince their national authorities to support them financially on the strength of the scientific quality and international significance of their joint endeavors. At the planning level, they contributed to the best of their abilities, reaching the apex with the formulation of the Horizon 2000 long-term program. Through their determination in defending what they saw as their common European program, the scientists forced and modified political decisions in favor of that program, which could not be taken for granted.

This intimate relationship between ESA and the scientific community worked, first of all, through advisory bodies made up of scientists representing the highest level of competence, bodies on which individuals served for limited terms and were replaced on a regular basis in order to allow maximum mixing between the scientific community and the ESA Executive. Initially set up only within the domain of space science, these advisory bodies quickly spread through all the Agency's science-related fields of activity. Their number, composition, and rules of procedure have changed with time as dictated by experience. At present the advisory bodies in space science are the Space Science Advisory Committee (SSAC) and its two related working groups, the Astronomy Working Group and the Solar System Working Group, which advise the Director General and the director of the scientific program on all scientific matters. In 1993 a special ad hoc working group in fundamental physics and general relativity was formed to advise ESA on the selection of the STEP project, a satellite to test the equivalence principle.

The Space Science Advisory Committee

The SSAC is composed of a limited number of senior scientists—seven to nine. The minimum number has less to do with the seven pillars of wisdom of T. E. Lawrence than with some subtle political alchemy. In fact, it is an accepted rule that the four larger countries should be represented on the SSAC. Scientists from the smaller nations are of course also included, but the total number is kept as low as practical so as to avoid the temptation to have as many committee members as there are Member States. The members of the SSAC are nominated by the Director General for a three-year period, after consultation with the SSAC itself and having heard the comments of the SPC. Two of the members are nominated as chairmen of the Astronomy and Solar System Working Groups. The chairman of the SSAC is nominated for a three-year period independent of any previous tenure within the SSAC. The chairman of the SSAC attends SPC meetings *de jure,* reporting on the activities of the Working Groups and presenting the recommendations of his committee. To maximize the exchange of information, the chairman of the SPC also attends *de jure* the meetings of the SSAC.

This close interlinking of the SPC, the SSAC, and the Executive came about as a compromise after a somewhat heated argument in the 1970s between the SPC and the Director General at the time, Roy Gibson. The SPC, wishing to receive advice directly from scientists rather than filtered through the Director General, expressed the intention to set up its own scientific panel of advisers. That was not favored by the Executive in view of the potential for conflicts between these two scientific bodies, one (the SSAC) advising the Director General and the other advising the delegates. The possibility of conflict was far from remote owing to the different briefs of these two groups: the SSAC, composed of members selected for their scientific competence, being responsible for defining a European policy consistent with, but not subservient to, national priorities; and the proposed panel, composed of members nominated by the national delegations, being more inclined to stick to national priorities and interests.

Gibson made it clear that, owing to the mandatory character of the Scientific Program and hence its intrinsic supranationality, it was vital that it be advised by the best scientists in the field. He offered, however, to inform the SPC in advance of his intended nominations to SSAC membership, and he accepted that delegations could make comments and express their views. In addition, the SSAC would be able to brief the SPC directly on its recommendations and analysis, in parallel to its reporting to the Director General. The SPC yielded. The recommendations issued by the SSAC are indeed directed in parallel and independently to the Director General and to the SPC. In case the Director General disagrees, and does not accept the SSAC recommendation, the Executive will inform the SPC, indicating his reasons and the essence of his different policy.

The Working Groups

The Working Groups are composed of scientists—not more than fifteen—and are meant to make expert recommendations in their respective fields of competence. Their composition reflects the various disciplines in the area of astronomy and Solar System exploration. Membership is for three years. The selection of members follows a rather peculiar rule which has had very beneficial effects. In general one-third of the members are renewed every year. The nominations are made, for 50 percent of the vacancies, directly by the members of the Group, without the Executive being able to interfere. For the other half of the vacancies, the Working Groups present a longer list of candidates from among which the Executive may—but is not required to—fill the positions.

This 50 percent rule was introduced in the 1970s and was branded in some quarters as an unacceptable interference from the Executive in the free recruitment of members. This even led to some resignations. In reality, the Executive's intention was to counter any creeping monopolistic impulses by appointing scientists from less privileged institutions, or from Member States in an evolutionary phase, to take part in the activities of the Working Group.

As in the case of the SSAC, the membership of the Working Groups does not have to reflect the relative weights of nationalities. However, a certain equilibrium is sought, which is the responsibility of the respective chairmen. Because of the fundamental role that the Working Groups and the SSAC play in the formulation of the planning and in the decisionmaking of the Agency, some countries try to exert, through their scientists, a decisive influence. These attempts have been largely unsuccessful, the scientists being too jealous of their scientific independence and too conscious of their European brief.

The Working Groups issue their recommendations to the SSAC, whose chairman attends their meetings. The chairmen of the Working Groups being members of the SSAC, the flow of information is thus optimized. The Working Groups' operating procedures follow those of the SSAC, with debates being followed by a search for consensus on the recommendations. More often than at the SSAC, when consensus cannot be reached, a vote is taken. Votes by proxy are not accepted, as those absent have not had the benefit of the various arguments raised in the course of the discussions.

One common rule for the SSAC and the Working Groups is that their scientists get no remuneration for the work they do, apart from reimbursement of their real expenses. The distinction associated with the membership is so great, however, that there has never been a scarcity of candidates.

Temporary Advisory Bodies

Apart from the standing advisory structure just described, the Executive may set up temporary committees or ad hoc working groups, to advise on particular subjects, which are disbanded after the completion of their particular task. Their common feature is that they are all composed of independent scientists, called to the groups for their particular expertise, independently of any consideration of nationality.

These groups either deal with particular aspects of a project or with problems of a more general nature. Typical of the first type are groups which are set up to select the payload complement of a newly approved project. The selection of the instruments is made competitively among proposals sent to ESA in response to a Call for Experiment Proposals. Also of this type are the groups which advise on the formulation of the observing program of an astronomical satellite used as an observatory by a broad community of users. These groups include scientists who are particularly distinguished in the field and who, so far as possible, have no conflicting interests—although the latter criterion is sometimes very difficult to meet. Often, non-European scientists are invited to participate in these groups. The Executive lends only technical support, and the choice is made entirely freely by the scientists. The recommendations of the group may not necessarily be in line with the interests or the views of the national delegations, who have to finance the selected payload elements and who must take into consideration the constraints of their budgets. Sometimes hard bargaining is involved, but delegations have, most of the time, been very cooperative when faced with choices made on a scientific basis, and this complex but open procedure has led to payload selections that have, by and large, met the consensus of the community.

Quite different from this category are committees which are called into existence for solving problems of a more general nature. Among these belong the committee on reciprocity, discussed later in Chapter 4, and the committee on the harmonization of ESA's policy for space science, microgravity, and Earth sciences, which was called upon in 1985 by Reimar Lüst, then Director General, and was chaired by René Pellat from France. That committee met in 1985 and 1986 and recommended that microgravity and Earth sciences, which are part of ESA's optional programs, should follow procedures similar to those adopted in the Science Program, in particular concerning the selection of missions and payloads and the formulation of long-term objectives.

Another ad hoc committee of a general nature was the so-called Survey Committee, which formulated the Horizon 2000 program, to which the scientific community of Europe contributed in a typical case of joint work toward a common goal which transcended all national boundaries. Because of its determining influence on Europe's space science activities, the work accomplished by this committee deserves more detailed discussion here.

▪ HORIZON 2000

ESRO and ESA had never had a long-term plan in space science approved and budgeted as such. It was argued by many, at all levels, that the fixed annual budget of the Scientific Program made long-term planning a futile exercise. Unless the global budget was increased annually, there could be no such plan. For fifteen years in the 1970s and mid-1980s the program enjoyed, but also to some extent suffered, a constant but rigid level of financing, with no decreases but also no increases. Within this tight constraint it had to cope with the explosion of new scientific areas.

Attempts at increasing the budget level were of course made, but they never succeeded because the case was not cogent enough, the arguments raised being mostly of a comparative nature with respect to other agencies or to other branches of science. National delegates would not even attempt to achieve the required unanimity to change the budget level unless there was a plan in front of them which would show what would be bought with the added funds. The challenge confronting ESA was to demonstrate the possibility of developing an evolutionary program, one with high scientific significance for Europe and for the overall space effort, that could be realized starting from the present budget level and progressing through realistic increases of that level, and completed in a time frame of about twenty years.

The authors of this book became deeply involved in this effort. Roger M. Bonnet was an outsider coming freshly into ESA and well aware of the needs and the requirements of the scientific community. Vittorio Manno had been operating under the system in place and knew its deficiencies well. Erik Quitsgaard, the Director General, fully endorsed the exercise, which became part of the preparation of the Agency's future long-term plan. The fundamental precondition of success was that the scientific community itself be engaged in the process of analyzing, reducing, and finally formulating a realistic minimum balanced long-term program.

Clearly, this called for a larger structure than that represented by the standing advisory bodies. The Survey Committee was then set up, of which the SSAC formed the core. The chairman of the SSAC, at that time Johan Bleeker of the Netherlands, took the chair of the new committee in order to ensure complete compatibility between the two bodies. The membership, in addition to that of the SSAC, included representatives from international organizations related to ESA's scientific disciplines: the European Science Foundation, CERN, the European Southern Observatory, and the International Astronomical Union.

Two scientific teams were set up in the areas of Solar System exploration and astrophysics. They were composed of representatives of all related space disciplines and maintained a continuing contact with the community at large. The first input from the scientific community was solicited in October 1983 in a letter from Bonnet calling for concepts for new missions to be undertaken over the next twenty years. The responses resulted in a list of more than seventy concepts which, together with a number of proposals already under consideration, constituted a very solid basis from which to work. Characteristically, the inputs were equally divided between astrophysics and Solar System science, accurately reflecting the interests—and the equilibrium—of the European space science community.

The teams separately focused on the difficult task of scrutinizing the concepts, and of deriving from them the contents of the future long-term program in each domain of interest. This required considerable organizational work on the part of the chairmen and of the ESA staff seconded to this task, in particular the secretaries of the Astronomy and Solar System Working Groups, Henk Olthof and George Haskell; the head of the ESA Space Science Department, Edgar Page; the head of the Science Projects Department, Maurice Delahais; and the head of the Future Program Study Office, Gordon Whitcomb. Countless meetings and consultations were held during a period of six months, at which the scientists of the teams debated among themselves, with the conviction that they were working for the best interest of all the community, creating a new and exciting program, and opening new and challenging perspectives for research.

This effort came to an end with a full presentation of the work of the teams to the Survey Committee which met in June 1984 during three days in the Fondazione Cini on the Isola di San Giorgio Maggiore in Venice. No site would have been more appropriate than this old convent, with the cloisters of Palladio and Longhena inspiring the major effort of synthesis on which the committee was about to embark.

The committee had not only to scrutinize and discuss the work of the teams but to distill it into an overall plan, a plan that had to be ambitious enough to meet the expectations of the scientists and to reflect the scientific tradition of Europe, while being framed within realistic financial and time limits in order to be acceptable to the political and financial authorities of the Member States. The debates, under the chairmanship of Johan Bleeker, were spirited and thorough. The intensity and concentration of the effort by all of the committee members could be gauged in some way by the fact that the telephone went completely unused, a rare event when so many heads of institutes were assembled. This focused brainstorming yielded Horizon 2000, ESA's first long-term plan in space science (see Figure 6). It was then the turn of the politicians and the administrators. Erik Quitsgaard fully endorsed the conclusions and supported them throughout the preparation of the ministerial Council meeting to be held in Rome at the beginning of 1985.

Surprisingly, the unanimity required to approve the budget for the program—which had been estimated to require a progressive yearly increase of about 5 to 7 percent over a period of ten years—rather than weakening the probability of approval, added to its strength by forcing into the limelight the program's coherence. The scientific community started working to persuade the national authorities to accept the program in toto, and to finance it. The requirement of unanimity for the Council to approve any change in the budget of the Mandatory Program led, after some debates, to the requested consensus.

At the January 1985 ministerial Council meeting in Rome, the Member States agreed unanimously to raise the science budget yearly by 5 percent above inflation for five years, and subsequently, at the December 1988 Council meeting, they agreed to continue for another five years—although this agreement was more difficult to obtain because of budget constraints in the United Kingdom. The coherence and equilibrium of the program, and the requirement of unanimity in the Council, presented recalcitrant nations with a black-or-white situation, and the cohesion of the scientific community achieved the rest.

It is worth stressing that Horizon 2000 is not simply a document, like so many in the past in Europe and elsewhere, amalgamating the wishes of those who prepared it and offering a shopping list of priorities. It is a real program, accepted unanimously by the thirteen Member States of ESA, which has been implemented systematically and without delay. That program has managed to raise the budget of ESA's Science Program by more than 55 percent over ten years, from a level that had been constant since 1970 and at a time when financial constraints affected nearly all research activities. It created more flight opportunities for the European scientists than ever before. It can be considered a real success, based on its support by the scientific community and on its realism in matching scientific ambitions to the financial realities of the Member States, through a somewhat modest but regular expansion. Very few planning exercises of this kind, in Europe or elsewhere, have been so successful. The Appendix gives a brief description of the content of this program.

One may wonder why a program similar to Horizon 2000 was not established in the domains of Earth observations and of microgravity. Earth and environmental sciences attract a lot of support from the politicians and governments in nearly every European country. Hence, one could easily imagine what effect the existence of a reference program could have in this area, in creating a set of key missions for the defense of our environment and the study of the global climate change. Europe could become one of the most important partners in this field.

Such a program is, however, more difficult to establish than in space science. One reason is that in ESA the Earth observation program is optional and bound to be more influenced by political considerations than the Mandatory Program. Earth observations are indeed at the borderline of applications; nevertheless,

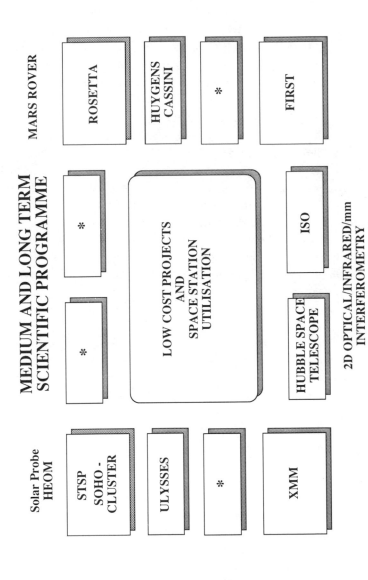

Figure 6. The structure of the Horizon 2000 program.

a large proportion of the Earth observation program is still of a purely scientific nature and should be treated accordingly. Scientists in the Member States have complained about the lack of a Horizon 2000 in Earth sciences and also in microgravity. The ministers, who met in Granada in November 1992, have recognized these defects and have requested the Director General to make proposals in 1995 to correct this unsatisfactory situation.

▪ NO SECRETS, NO SURPRISES

In principle, according to the Convention, it is the Council which fixes the overall space policy and which defines the long-term objectives of the Agency. History tends to show, however, that these main orientations are decided at the ministerial Council meetings, while the medium-term objectives, the program plans, and the budgets remain under the responsibility of ordinary Council meetings. The implementation of the programs is the responsibility of the Executive. The Council, the SPC for the Science Program, the Administrative and Finance Committee (AFC), the Industrial Policy Committee (IPC), and the different Program Boards in the case of optional programs, exert external control on the Executive. These committees induce a natural spirit of openness, but, at the same time, make ESA more vulnerable to criticism.

The set of ESA's delegate bodies, as of 1993, is illustrated in Figure 7. These bodies are made up of delegates representing the Member States and, depending on the relevant agreements, of associate or cooperating states. They make decisions or recommendations intended for all the governments or for the Director General.

One day, in the course of an unofficial discussion with Bonnet, Jean-Marie Luton, then the French delegate at the Council, raised the issue of the election of the new Director General. At that time, the name of another and very enterprising delegate was in everybody's mind for the post. Luton expressed some doubt that this choice would eventually be made. His rationale was, as he said, that ''there is no way you can make a coup in ESA,'' and that, according to him, the delegate was obviously not the kind of person to accept that. Indeed, the delegate never applied, and another one was eventually elected: Luton himself became the Director General of ESA on 1 October 1990.

The openness of the ESA's procedures makes surprise unlikely. Program decisions are known long before they come to the stage of implementation. This is the logical consequence of the process through which new projects are

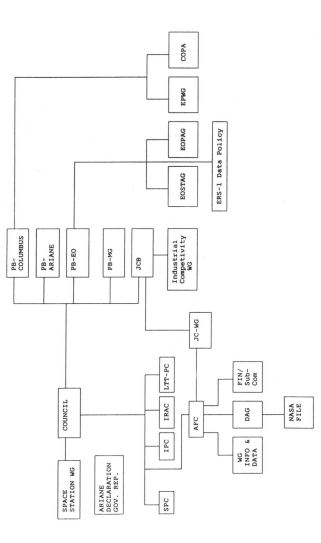

Figure 7. The structure of ESA's delegate bodies. AFC = Administration and Finance Committee; COPA = Columbus Payload Working Group; DAG = Documentation Advisory Group; EOPAG = ERS-Operations Planning Advisory Group; EOSTAG = Earth Observation Scientific and Technical Advisory Group; EPWG = Eureca Payload Working Group; FIN = Finance Sub-Committee; IPC = Industrial Policy Committee; IRAC (or IRC) = International Relations Advisory Committee; JCB = Joint Board on Communications Satellite Programs; JC-WG = Joint Working Group (AFC/JCB) on Commercialization and, in particular, DRS Commercialization; LTT-PC = Long-Term Space Transportation Systems Preparatory Program Committee; NASA = NASA File ad-hoc Group; PB-Ariane = Ariane Program Board; PB-Columbus = Columbus Program Board; PB-EO = Earth Observation Program Board; PB-MG = Microgravity Program Board; SPC = Scientific Program Committee; WG INFO & DATA = AFC Working Group for Rules on Information and Data.

introduced in the Agency's program. At this point a distinction is again necessary between the mandatory and the optional programs.

The Mandatory Program

In space science, we have seen that the initiation of the long-term plan and new missions is completely in the hands of the scientific community. Such an open and broad consultation leaves no room for secrets. From the first call for ideas to the screening by the various advisory bodies and the involvement of the science teams advising the Executive during the various phases, through the final decision by the SPC, the broadest information circulates, and it is difficult if not impossible to create any surprise. ESA has no rule of confidentiality whatsoever that would prevent the announcement of the selection of a scientific payload prior to the completion of all the phases of the selection process.

The "unexpected" may nevertheless occur at some stages. For instance, it may well happen that the SSAC does not follow the priorities recommended by the working groups. This has occurred at least twice in the history of ESA. The first time was in 1978 when a comet mission—which eventually became Giotto—was retained by the SSAC against the advice of the Solar System Working Group, which favored a Lunar Polar Orbiter. The second time was in 1991, on the occasion of the selection of the Phase A candidates for the second medium-sized mission of Horizon 2000. The SSAC gave a lower priority to IVS, a very long baseline interferometry mission working at radio wavelengths, to the benefit of PRISMA, a satellite devoted to stellar seismology, which the Astronomy Working Group had ranked second after IVS. In each case, the change in priorities was justified. But, because the situation was unusual, it created some surprise—and disappointment—in the community of planetologists and radio-astronomers.

The SPC also created something of a surprise when, in March 1980, it inverted the priorities assigned by the SSAC and gave preference to Hipparcos, the astrometry mission,[1] over Giotto. Giotto was selected a few months later. The success of both missions is an indication of how difficult was the choice and how justified was the selection of both.

At all stages in the implementation process, the Executive keeps the SPC informed of progress and of problems encountered. After the approval of Phase B, the technical and financial evolution of the approved project is reviewed at least three times a year, the SPC being entitled to stop it if its budget overruns its approved cost-at-completion by more than 10 percent (this limit is set at

20 percent for the optional programs). Such a situation is very rare, and even when it has occurred, at no time has the SPC decided to end the project.

The Optional Programs

The case of optional programs leaves even less room for surprise. The Agency's optional activities are authorized and controlled by the Council and monitored by the delegations in the arena of the Program Boards, in addition to the AFC, which has to verify the compatibility of all actions with ESA's administrative procedures, and the IPC, which verifies the correct application of the industrial policy. On special occasions, expert groups are created as well as special ad hoc groups and joint groups. This structure may appear confusing to the uninitiated; indeed, even those familiar with it occasionally wonder where the authority of one committee stops and that of another begins.

New optional programs can originate either from the Executive or from the delegations themselves. For instance, the Ariane and Hermes programs were initiated by the French delegation, and this came as no surprise: the intention of the French was publicly known long before Ariane was proposed to the Council for "Europeanization." The next step is to set up a meeting of "potential participants," made up of the delegations that are seriously considering investing money in the future program. There the program proposal is examined, each delegation focusing on the potential return to their industry, and the legal instruments that will govern the management of the new candidate are drafted. These meetings are often chaired by the Executive. In principle, the management of programs is the responsibility of the Executive, but there is a tendency for the relevant Program Boards to be more involved in the daily running of the projects.

Through the Council, the committees, and the Program Boards, how could a secret be kept? In fact, only the restricted meetings of the Council are subject to strict confidentiality. For the "normal" meetings, the decisions are known openly. Jokingly, ESA staff members like to say that if you want to know what is going on in the Agency you should ask the NASA representative in Paris: he knows everything, faster than anybody else, and usually quite accurately. Conversely, NASA says that the ESA representative in Washington knows more about NASA than the NASA staff themselves!

More seriously, an international agency like ESA has to be open and has to deliver information. It is controlled very carefully by the delegates and by all the users it is supposed to serve. How could it be otherwise? ESA's record of success testifies that this policy is not a handicap, although it may make ESA more open to criticism and more difficult to manage, and may occasionally discourage some enterprising candidates for the post of Director General to apply.

▪ IS THERE A LANGUAGE BARRIER?

Misunderstandings among the scientists and the engineers, errors of interpretation, legal issues arising because of the difficulty of translating idioms—these are potential handicaps which may affect an international organization such as ESA. A workforce of people coming from thirteen different countries may indeed present some communication difficulties and create problems for the top management.

Scientists, by tradition and by necessity as well, usually have to express themselves in English. For them, the problem of a language barrier is nonexistent, which does not necessarily mean that they express themselves well in the language of Shakespeare. Engineers and administrators, on the contrary, are not necessarily used to using a language that is not their mother tongue, although at ESA a large majority of them do, and sometimes better than the scientists. They bear the responsibility for building sophisticated hardware, sometimes with many complex interfaces, or of negotiating contracts and memorandums of understanding, which should leave as little room as possible for ambiguous interpretations.

Looking at it, it does seem that there is no such thing as a language barrier in ESA, and this is reassuring. There are, however, a few subtle problems which make the daily life there sometimes trickier than if all the people involved shared one mother tongue. Often, it also makes life more interesting.

ESA has three official languages: English, French, and German. The Convention itself exists in seven languages. In practice, however, English and French are used as the two main working languages. The term ''official'' means that every delegation has the right to express itself in any one of these languages. It means also that Council and committee or Program Board meetings are assisted by interpreters in each of these three languages. Documents prepared by the Executive must be published in all of them. The tradition is that they are printed in three different colors: blue for English, white for French, and green for German. In everyday life, English, or some form of English, is generally used. In fact, two-thirds of the working papers are written by non-British staff. This does not make the life of the translation office all that easy, and sometimes the English texts are difficult to understand, even for the British themselves.

The fact that the natives of twelve of the thirteen Member States do not express themselves daily in their mother tongue places nearly everyone on an equal footing with respect to language, with, of course, the exception of the British—who, on the other hand, have to listen to 85 percent of the staff speaking some kind of ''broken English.'' This situation forces everyone to pay more attention to what is said during meetings and to be indulgent with the others. The predominance of English is of course a problem, and some ESA employees have more difficulty than others in using this language. At the time of recruitment, during interviews, knowledge of English is an important factor for selecting new staff members, depending on the job, although in principle the candidate has the choice of using either English or French. There is no doubt that a good command of the English language represents an asset. This kind of bias may explain why the British have for a long time been predominant in number at ESA, a situation which has forced the Administration to apply quotas for the different nationalities. These quotas tend to reflect the percentage of Member States' contributions (Table 5).

The near obligation to speak English does not hold for the delegates. In fact, French delegates usually speak French, the British English, and the Germans German. All the others express themselves in any of these languages. This at times does create some interesting, if not delicate, situations. The SPC learned to differentiate between the interventions of one delegate who used English on scientific matters and one of the other two official languages on political or administrative matters. No one in ESA will ever forget that late distinguished Spanish delegate, General San Aranjuez, who used French, but whose accent was so pronounced that even the French delegation had difficulty understanding him. So much, in fact, that they had to listen to the English interpreter who could understand and translate the general's language. Many times, the general was obliged to vote negatively or to abstain, with one memorable exception when he expressed himself very clearly and positively, leaving no room for any ambiguity. This was on the occasion of voting on the continuation of the operations of the joint ESA-UK-NASA IUE spacecraft. The IUE telemetry station is located at the ESA station at Villafranca near Madrid, and its presence does represent a real asset for the Spaniards. Not only does it provide a visible image of the involvement of their country in the European space program, but it also offers a training opportunity for Spanish astronomy students. As is usually the case at the time of voting, the SPC chairman, at that time Kees de Jager, made a *tour de table,* asking every delegate to take a stance on the matter. When it came to Spain, the general was surprisingly beaming. He raised both hands and loudly said in perfect English, ''Yes! Yes!'' This was indeed a clear vote.

Table 5.   Distribution of ESA staff by nationality at the end of 1992

Country	Number of staff	% of total staff	Budget contribution as % of total budget
Austria	28	1.4	1.1
Belgium	99	4.8	5.1
Denmark	37	1.8	1.0
France	445	21.6	30.6[a]
Germany	402	19.5	24.4
Ireland	20	1.0	0.2
Italy	288	14.0	17.2
Netherlands	240	11.7	2.6
Norway	20	1.0	0.8
Spain	91	4.4	4.8
Sweden	56	2.7	2.5
Switzerland	29	1.4	2.5
United Kingdom	271	13.2	6.3
Other countries	31	1.5	1.1
Total	2,057		

a. The large difference between France's percentages of contribution and of staff is due to the Ariane project, which is subcontracted to CNES and for which ESA has almost no staff.

Such are the daily problems of a multilingual organization which, in fact, has faced very few difficulties as a result of 85 percent of its staff expressing themselves daily in "broken English." In reality, it is sometimes more difficult for the uninitiated to get the substance of a paper, or of a conversation, because of the overuse of incomprehensible acronyms. But this is not true only at ESA.

■ INDUSTRIAL POLICY

The science policy is to serve the scientific community, independently of political interests. The industrial policy is to serve the industrial, technological, and very often the political interests of the Member States. These two policy ingredients, in spite of their apparently diverging objectives, have played a major role in ESA's overall success. In fact, ESA has no capacity to build the equipment for its own missions, which is, for the most part, subcontracted to European industry: about 90 percent of ESA's budget is spent on external contracts.

If the original motivations of the founding fathers of ESRO were predominantly of a scientific nature, the ministers present at the conference of plenipotentiaries in June 1962, when the ESRO Convention was opened to signature, clearly expressed their concern that space activities should preserve and contribute to the development of the European industrial potential. In accordance with this position, they issued the following resolution: "The Conference recommends that the Organization shall place orders for equipment and industrial contracts among the Member States as equitably as possible, taking into account scientific, technological, economic and geographical considerations." This recommendation is one among those which the ESRO Council adopted at its first meeting. It is the basis on which the industrial policy of ESRO was established, and it has been applied also in ESA, where it has been considerably developed and refined.

The principles of this policy, explicitly described in the Convention, can be summarized as follows: in the execution of the Agency's programs, maximum use should be made of industry, as opposed to building up an ESA in-house capability; utilization of industry should be based on free competitive bidding; developments and procurements should in principle make use of Member States' industries; an adequate geographical distribution of contracts to Member States' industries should be ensured; a balanced development of a competent European space industry should be aimed at through the use, structuring, and rationalizing of existing industrial capabilities; measures should be taken to improve the worldwide competitiveness of the European space industry; launchers developed by the Agency should, in principle, be used in support of European programs.

This policy applies only to the systems developed by ESA that are the common facilities offered to the users—the spacecraft, the ground stations, the infrastructure, the operations of the satellites once in orbit. It excludes the development of scientific instruments that, under the responsibility of the experimenters, are funded by particular nations.

Definitions

The proper implementation of these principles requires precise definitions. For example, one must be clear, when dealing with a company, whether or not it belongs to a Member State. There are obvious and clear cases, such as that of a company whose seat and manufacturing facilities are located in a Member State, or, on the other hand, the case of a company whose seat and manufacturing facilities are *not* in a Member State: even if such a company has a commercial office in a Member State, it is not considered to belong to that Member State. There are also more ambiguous situations which require careful analysis. For example, there may be a multinational company whose seat is not located in a Member State but which develops a large number of activities, including research and development, on the territory of that state. Or there may be a company whose executive center is located in a Member State but which rests largely on technologies, or equipment, or support, which are developed or produced in a non-member country. Or there may be a multinational company whose executive center is located in one Member State but whose technical support facilities are located in a different one. In each of these cases, a detailed analysis is necessary.

Use of Outside Companies

It often occurs that for technical or for schedule reasons, a mission has to rely on a company from outside the Member States.[2] It is admitted, however, that this must be done within certain rules. For example, for the sake of fairness, such firms are invited to bid on projects only if their chances of being selected are good. For research and development activities, consultation of firms from outside the Member States is excluded in principle, except in those cases where the envisaged technologies would take too long to develop in Europe. Invitations to tender bids can also be issued to outside firms if the Member States' offers are considered to be too expensive, or if their expected delivery date is incompatible with the schedule of the project.

In the case of scientific projects, which, to a large extent, are development projects, it is felt that European industry is generally competitive, except for subsystems for which U.S. industry, for instance, can rely on technologies or production series that are nonexistent in Europe. In summary, the rule is in general not to consult a firm from outside the Member States except when there is no other choice.

The Principle of Juste Retour

Of all the principles which govern ESA's industrial policy, the most famous is the industrial return principle, also known as the principle of *juste retour*.[3] The implementation of this principle is measured through the industrial return coefficient, defined as the ratio of the portion of all contracts placed by the Agency in the industry of a given Member State to the average percentage of contributions of that Member State to the budget that the Agency spends in industry. The return is industrial and not financial in nature. For this reason it is weighted according to the technological interest of the various activities.

Weighting factors can vary between zero and one, the highest being attributed to most of the flight hardware and the lowest generally to ground support equipment. For instance, mechanical ground support equipment is weighted 50 percent, while launch services, although representing a considerable part of a project cost, are weighted only 25 percent. The non-weighted part lies between 50 percent and 80 percent of all the industrial contracts. The *juste retour* is a global notion that encompasses a Member State's participation in all the activities of the Agency, whether mandatory or optional, since 1972. It is measured in accounting units. The computation is first made program by program and then cumulated to all the programs.

Space programs are powerful means of promoting technological development. Space expenditures in ESRO and ESA have increased by more than an order of magnitude since the ESRO days and, as a consequence, space programs occupy an increasing portion of the activities of the aerospace companies in Europe, which have invested a considerable amount of money and effort in order to remain competitive. European aerospace industry involves more than 38,000 employees in Europe, distributed over nearly one hundred companies. It has to operate in an environment marked by ever-fiercer competition. Both Member States and Executive, therefore, have an interest in the implementation of the industrial policy, and it should come as no surprise that the Member States insist on getting their share of this market. The industrial return coefficient is one simple and visible way of measuring their involvement, and they are keen to keep it at the highest level, if possible of course above 100 percent, an obviously impossible target to reach for all of them at the same time.

The implementation of *juste retour* has evolved considerably since the early days. In ESRO's day (1967) the Director General was requested to ensure a minimum return of 70 percent to all the Member States. This percentage was changed in the ESA Convention, which authorizes the Director General to take exceptional measures and requests him to inform the Council if the return coefficient of a Member State is lower than 80 percent, in order to redress the situation within one year. At the 1985 ministerial meeting in Rome, the Director General was requested to take measures to ensure that the cumulated coefficients for all Member States be not lower than 95 percent at the end of a three-year period ending in 1987. The limit below which special corrective measures had to be taken was then raised from 80 to 90 percent. At the 1987 ministerial meeting in The Hague, the rules were changed once more: the 95 percent minimum coefficient target was confirmed, and all participating Member States were guaranteed a minimum return of 90 percent per program, together with compensation for the cumulative deficits since 1972. This topic was again discussed in Munich in 1991 and in Granada in 1992, where the 90 percent limit was raised first to 95 percent until 1993 and then to 96 percent for the three-year period 1994–1996, it being understood that the objective of the policy should be for every Member State to have a return coefficient as close as possible to 100 percent.

As we can see, the rules of industrial return have become more precise and restrictive, and its implementation represents an increasing concern for the Member States and for the Executive. It is indeed remarkable that ESA manages its programs within these constraints, while at the same time meeting return coefficients which are not far from the ideal target.

Implementation

The committee in charge of following the implementation of the industrial policy is the Industrial Policy Committee (IPC), at which the delegations discuss the implementation of the policy for all major contracts and especially when a project goes through Phase B or C/D in industry (see Figure 8).[4] If the delegates disagree with the proposal of the Executive they can vote against it. A simple majority is sufficient to get a contract approved.

When a new project starts, return targets are established by the Executive for each Member State, in order to compensate either for existing or potential deficits or for surpluses. After the selection of the prime contractor (see Figure 9),[5] usually at the start of Phase B, a procurement plan is agreed upon, which should in principle meet the geographical industrial targets while achieving, through as much competition as possible, an acceptable estimated cost-at-completion. Phase B is of crucial importance for ensuring that both the targets and the cost limitations are respected.

It is usually the role of the prime contractor to form an industrial team and to make sure that it obeys the rules fixed by the Agency—although in some instances the Agency takes over this role. During Phase B, the status of the project is closely monitored by a joint Agency/prime contractor procurement board. At the end of Phase B, at the transition to Phase C/D, the IPC can check and vote on the proposal of the Executive based on the work done by the prime contractor.

At the end of the exercise, deviations from the ideal target do indeed exist, but if the work has been done properly during Phase B, in other words if there has been enough dialogue between the Executive and the prime contractor on the one hand, and between the Executive and the delegations on the other,

LIFE CYCLE OF A TYPICAL SCIENTIFIC PROJECT

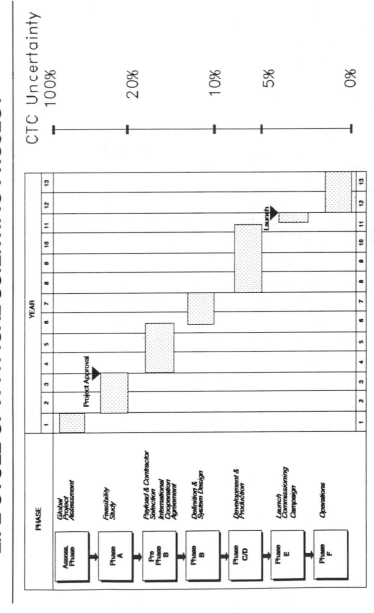

Figure 8. The succession and timing of the various phases of development of a project. The margin of uncertainty on the cost-at-completion is shown on the right.

these deviations are few and the proposal of the Executive is usually accepted without too much difficulty by the IPC. When a project has slipped or has faced technical difficulties, the actual return coefficient may deviate from the ideal original target. Most often, this results in surpluses for the country to which the prime contractor belongs—one of the larger contributing Member States—and, correspondingly, to deficits in some of the smaller countries. In the case of the Science Program, which is mandatory and ensured of continuity for as long as the Agency itself continues to exist, any deviation encountered in a project can in principle be corrected in the implementation of the next one. It is easier to meet the targets in the case of the bigger projects, for example the cornerstones of Horizon 2000—STSP, XMM, Rosetta, and FIRST—which rest on bigger budgets, than for smaller or medium-sized missions.

This juggling procedure is complex, delicate, and not easy to implement. It is worth analyzing how it has worked, and whether it has fulfilled any useful role in the overall ESA management as well as in Europe and its aerospace industry in general.

Results

In spite of its apparent complexity, the principle of *juste retour* has had a clearly positive effect on the determination of the various Member States to support an international European space program and to participate fully in the Agency's activities. It has also played a determining role in the overall increase of these activities in Europe, as seen in comparison with other research and development programs: between 1983 and 1987 the civilian research budgets of the largest contributing Member States—France, Germany, Italy, the United Kingdom, Belgium, and Spain—increased by 8 percent on average, while their corresponding civilian space budgets increased by 60 percent. This progression would not have been possible without the industrial and the technological asset that space activities represent, and the assurance, mostly for the smaller Member States, of a fair return of industrial contracts.

In fact, the existence and development of a space industry in the smaller Member States rests essentially on their certainty of getting a minimum level of activities, as secured by the *juste retour* principle. To fulfill this principle, the Agency has had to look for a diversity of competencies in the various countries, particularly in the smaller ones, where the existing capacity was not always sufficient to ensure a return compatible with their level of contributions. This enlarged spectrum of industrial capabilities, and the *juste retour* principle,

PRE-PHASE B CONTRACTOR SELECTION

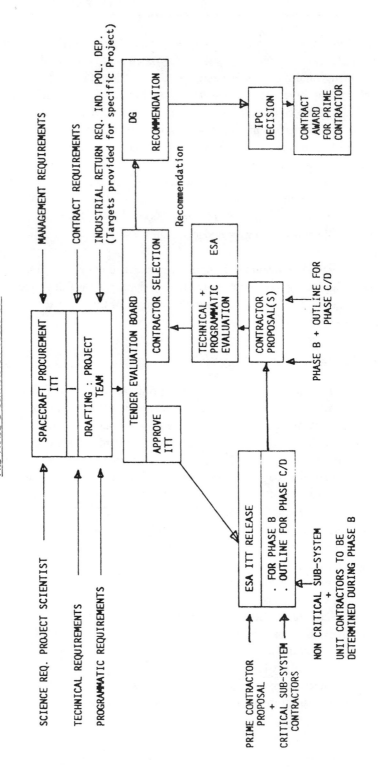

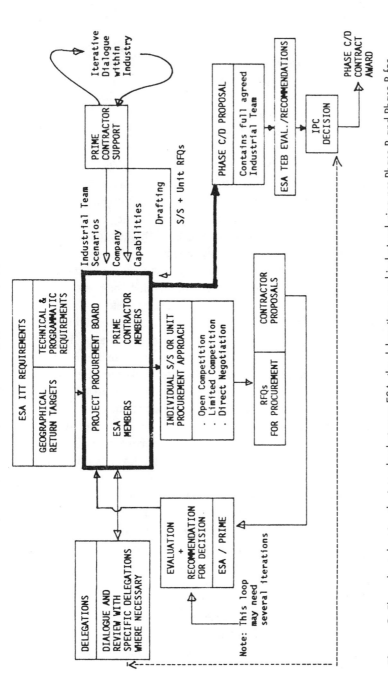

Figure 9. Flow chart showing the interplay between ESA, the delegations, and industry during pre-Phase B and Phase B for scientific projects. In order to secure a fair return target for each Member State, it is necessary to go through a complex process which involves a long series of iterations between the Executive (project scientist and project manager), the delegations, in particular at the IPC level, and the prime contractor (selected in competition prior to the start of Phase B) and its subcontractors during Phase B. Subcontractors can be selected through either open competition, limited competition, or direct negotiations. The award of the final contracts for Phase C/D is done through the IPC.

have reduced the risk of monopolistic situations which might have led to abuses. In the absence of the *juste retour* principle, space activities in Europe would probably have been concentrated in a set of, at most, a dozen industrial firms, as compared with the one hundred currently involved. Such a situation would have made it difficult for many Member States to continue participating in the ESA programs, and would have jeopardized the very existence of the European space effort at the level it has reached today.

Is this industrial policy cost-effective? This question has no precise answer. There is little doubt that in the context of this policy, the management of large programs, involving sometimes more than forty different firms in more than thirteen countries, induces additional costs and adds a certain degree of complexity and therefore of inefficiency, as compared to a situation where the same program would be developed by a minimum number of firms distributed in only one or two countries. The "atomization" of subcontracts increases the number and the complexity of interfaces and naturally also the costs of management, forcing the inclusion of risk margins for deficiencies.

However, in spite of the large number of projects already undertaken by the Agency, there is no easy way to quantify the cost increase induced by the application of the *juste retour* principle. There are nearly as many individual cases as there are individual projects. Inasmuch as the firms compete, one can say that the influence on cost is minimized. This is more true in the case of scientific projects for which competition, in principle, applies at all levels, including the choice of the prime contractor; optional programs are often initiated on the understanding that the most interested Member State will also get the prime contract, thereby restricting competition to the level of subcontractors.[6]

At this stage, we may say that ESA's industrial policy has been an essential factor in the overall success of the Agency. Certainly, there is a price to pay, an increase in the cost of projects as compared with what they would be in a totally free competitive context, although there is no easy quantification of this effect. In addition to the active involvement and cooperation of the scientists, the integration of a large number of industrial companies in the development of space projects under the leadership of a prime contractor has been an important element in the overall effort of European integration, and space industry is certainly one of the sectors where a true "European spirit" prevails. It is possible that the recent opening of European borders and the implementation of the Single Act by EEC countries may affect ESA's industrial policy. We defer the analysis of this aspect to Chapter 6.

In fact, the remarkable success of ESA's industrial policy, together with the success of its missions, does explain why both industry and the scientific community are strong supporters of ESA's space program. If the cost-at-completion of projects may increase as a side effect of the industrial policy, it may well be a price worth paying to allow Europe to undertake missions of a unique scope and character, with a prominent place among all the space projects undertaken around the world.

■ COMPARISONS WITH OTHER ORGANIZATIONS

NASA

ESA cannot directly be compared with NASA, the U.S. space agency. The comparison is nevertheless frequently made, and it is useful to clarify the differences between the two agencies. ESRO was not, and ESA is not, a one-government organization. ESA is multinational, governed by a Council on which every Member State is represented, and whose budget sums up their individual financial contributions. While both ESA and NASA do conduct space missions for peaceful purposes, ESA is not alone in this role in Europe: it coexists with national organizations such as the Centre National d'Etudes Spatiales (CNES) in France. In fact, CNES can be more directly compared with NASA than ESA can. CNES is responsible for the French space program and NASA for the U.S. space program.

More subtle differences do exist in the management of the respective scientific programs of ESA and NASA, some of which originated in the early times of ESRO. These are, for example, the distribution of responsibilities between the Member States and the Agency, and the fact that work is mostly subcontracted to industry, with very little being done internally. ESA has no in-house capability to build the equipment for its missions, while most NASA centers do possess such a capability. (The only exception to this rule was the SLED, equipment to test the reactions of the human body under various levels of linear acceleration; it was developed at ESTEC and used successfully for the first time in 1985 by ESA astronaut Wubbo Ockels on board the D1 Spacelab flight.)

Very early, the European scientists, in particular those from the United Kingdom, wary of the tendency of space agencies toward excessive bureaucracy, took a very firm position against scientific work being carried out by ESRO staff; the scientists considered it their own prerogative to define and to review the objectives and the scientific content of ESRO missions. This bottom-up approach is also the basic principle governing the working of ESA's Scientific Program.

A closely related issue is that of financial support to scientific groups for the development and operation of their experiments and for the analysis of their data. Within the American system, NASA covers the costs of (national) experiments accepted on their satellites, whether they are proposed by scientists within NASA centers or by external (U.S.) groups. In order to preserve the independence of research groups in Europe, the founders of ESRO recommended that experiments proposed by scientists be financially supported from the national resources of their respective Member States. Only big facilities—the satellites, the launchers, the tracking and operations—would be financed from within the Agency budget. This approach has by and large been followed by ESRO and ESA, with two notable exceptions due to the fully integrated character of the missions: Hipparcos and the Faint Object Camera on board the Hubble Space Telescope.

These principles were introduced in the structure of ESRO when the question of the role and the size of ESLAB were debated. It was accepted that this research unit would be small, with a reduced permanent staff and a few research fellows. Its main role would be to ensure the scientific integrity of missions by providing project scientists who would, among other things, be responsible for maintaining the scientific value of the mission when confronted with conflicting technical and engineering requirements. In order to remain active in science work, the project scientists should be able to carry out research and to develop scientific experiments of their own. The Space Science Department of ESA, which is the remnant of this unit, today totals about sixty staff members. It is located in ESTEC and is headed by Martin Huber from Switzerland. There is no equivalent, either in Europe or within the ESA system, of a research/operational institute like the U.S. Jet Propulsion Laboratory, whose staff presently exceeds four thousand. This is another striking difference between the American and European space organizations.

CERN

Interestingly enough, although ESRO's organization was derived from the example of CERN, and although the two organizations had many of the same founding fathers, they have followed different paths. CERN was established primarily to conduct fundamental nuclear research in its own right, and not purely as a service organization. It is devoted to the single, though very demanding, scientific discipline of high energy physics. The kind of science conducted at CERN is of a fundamental character, with no immediate fore-seeable applications. The concerns of ESA, in contrast, are very diverse, covering many disciplines, from research on plasma and magnetospheric physics to the exploration of the Solar System, the study of planetary atmospheres, of the heliosphere, of the Sun, and of astrophysics. The development of satellites and automatic systems in space, and of the related technologies, represents a strong incentive for European industry. All CERN facilities are concentrated in a single big center near Geneva, and all its experiments are carried out from there. The concept of geographical distribution, which has played such an important role in ESRO and ESA, virtually does not exist at CERN, nor do considerations of industrial return strongly influence the management practices or determine the way contracts are distributed.

The structure of ESA is undoubtedly complex, but so is Europe. The coexistence of an international European effort and national programs would be inefficient without the coordination that is one of the prime mandates of ESA. Because the larger contributing Member States represent more than 75 percent of the overall budget, the Agency could probably survive without worrying too much about the smaller Member States, which, furthermore, present the most delicate industrial return problems. However, concentrating the European space effort within these large countries would be wrong. The smaller Member States are in fact strong supporters of ESA if not the strongest, since they rely mostly on the Agency to secure their programs, to foster their industrial capabilities, and to improve their technological level. They are the cement of the organization. It is of course unavoidable that the larger Member States play a major role in defining European space policy. The implementation of that policy, however, would not be optimal without the smaller Member States. This complementarity, together with the success of the missions and programs such as Ariane, have induced a true spirit of solidarity. It can be identified as the main reason why ESA has been such a successful European enterprise.

3

THE AGENCY AND ITS MEMBER STATES

ESA belongs to its Member States and is entrusted with the task of defining a space policy and of developing systems to be made available to the scientific community and to the users of space application programs, such as telecommunications, meteorology, and remote sensing. National space organizations coexist with ESA, interacting with it in diverse ways. Greater efficiency might be achieved through a progressive integration of all the European space activities, as claimed in the ESA Convention, but this goal seems to be very difficult to achieve and, in fact, not necessarily desirable. It remains, even after nearly thirty years of existence of ESRO and ESA, a very remote objective. The first step should be, more realistically, to coordinate the various national programs with the ESA program, a task that, even in the space science area, which is less dominated than others by political considerations, has been undertaken only recently, with the Horizon 2000 program.

▪ NATIONAL AND EUROPEAN PROGRAMS

An important differentiating element between the various Member States is the existence or not of a strong national space program and space organization. Another distinction is the proportion of their GNP that Member States commit to their space activities. Those with the lowest percentage show a tendency to rely more on the Agency than those with a higher percentage (Table 6). Historical considerations are often important as well. Initially, well before ESRO existed, a number of European states, in particular the United Kingdom, France, and Italy, embarked on space science activity of some kind. A number of national scientific satellites were developed and launched, such as UK-1 in

Table 6. Global statistical data and breakdown of the space expenditures of ESA Member States (in MAU).

Member State	GNP[a]	Total space expenditure	Total contribution to ESA	Space science[b]		No. of space scientists	Expenditure per scientist
				ESA	National		
Austria	121	30.2	27.00	6.2	2.4	32	0.27
Belgium[c]	146	126.6	113.4	7.05	6.71	37	0.38
Switzerland	182	64.07	55.79	10.67	5.5	49	0.57
Germany	1170	629.00	456.5	56.8	58.2	340	0.34
Denmark	96	30.6	25.6	4.9	1.03	20	0.30
Spain	374	191.5	113.5	15.8	4.37	49	0.41
Ireland	30	6.50	5.7	1.5	0.8	14	0.16
France	919	1535.00	672.00	46.6	61.6	400	0.19
United Kingdom	812	224.9	141.3	38.1	20.7	500	0.12
Italy[d]	860	753.00	373.00	38.2	31.2	450	0.15
Netherlands	218	88.7	64.9	12.00	5.5	70	0.25
Norway	86	52.00	19.00	4.5	1.95	40	0.16
Sweden	188	76.4	52.20	8.64	4.0	70	0.16
Finland	110	33.7	8.01	4.62	8.47	37	0.18
Total	5312	3842.17	2127.90	255.58	180.13	2108	0.20

Source: These data were provided by the ESA Member States at the 1992 Capri meeting.
a. 1990 data in Giga Accounting Units.
b. Space science data do not include Earth, microgravity, and life sciences.
c. This includes not only purely national activities but also the financing of payloads and the reduction and interpretation of the data.
d. 1991 data.

1962 (United Kingdom), San Marco-1 in 1964 (Italy), and FR-1 in 1965 (France). This list is by no means exhaustive. Other nations—Sweden, Spain, and Germany—had sounding rocket programs.

These early efforts were generally successful and were the seeds from which an integrated European effort was able to grow. It was apparent, however, that the magnitude of the effort necessary to develop space science in Europe to a level comparable with that of the United States or the USSR could not be afforded by any single state. For scientists throughout Europe to be able to participate in this new scientific discipline, an integrated international effort involving the best scientific talents, the best industrial capacity, and the financial means of the majority of the European states was the only way.

This integrated approach was certainly not meant to stop the national programs. Germany, Holland, and Sweden joined in the development of satellites of their own. Their undertakings were not intended to compete with ESRO, but rather to complement ESRO's program with projects which corresponded more closely to their own specific needs, as well as to develop the scientific and technological capabilities that would allow them to participate effectively in ESRO's and ESA's programs. This was not a problem as long as ESRO (the only space science agency *of* Europe, though not the only one *in* Europe) could continue to provide the major facilities required by the scientific community and to exercise the coordinating role which its Convention claimed.

This was certainly not an easy task, as some of these national programs had their proper dynamics and their own political motivations. Very often, close relationships were independently established and carried through with the United States or with the Soviet Union. The stronger its national program, the more insistent was a Member State to exert its influence on the ESA program. However, politics never did come to be the overriding factor in the definition of the Science Program. All scientific decisions of the SPC are based predominantly on scientific evaluation by scientists selected on the basis of their scientific knowledge before any consideration of flag comes in. Whenever national priorities have been able to influence ESA's program, it has been because they were scientifically correct and in line with the scientific aims of ESA.

On the other hand, those Member States which did not have their own national programs expected ESA to play the role of a national space agency for their scientists and their industry. The circumstances and needs were very different among these states. As an example, there were states which did not have an administrative framework for financing the experiments selected by

ESA on its satellites. Thus, a new program, Prodex, was created in 1986,[1] through which the Agency could offer its administrative structure for managing this problem. In some other cases, in particular when the absence of an appropriate infrastructure would put scientists and engineers in a Member State at serious disadvantage as compared to their colleagues from other European states, the Agency has helped with advice and, through its establishments, with some training.

Hence, the Agency's role was different depending on whether a state did or did not have a national program, and the circumstances also differed widely within each category. It was vital, however, to preserve its role as the Agency of Europe and, as such, capable of delivering systems which in principle were beyond the capabilities of any single Member State. In the 1980s, however, this role was challenged. The growing importance of space science in its interrelationship with other branches of science, and the increasing number of scientists involved, were demanding a larger frequency of flight opportunities. At the beginning of the 1980s, the scientific community in Europe consisted of some forty hardware-developing groups, as well as more than a hundred institutes involved in using space data, totaling some two thousand scientists and engineers. European industry was organized in three major consortia bidding competitively for the development of projects. By that time the envisaged facilities had become much more sophisticated, another sign of the rapidly evolving character of space science, and much more costly. European space scientists and industry requested a certain frequency of projects to guarantee continuity of research for the scientists and of workload and technical expertise for industry. Unfortunately, ESA's level scientific budget could not accommodate this double request. There was a risk that both the scientists and industry would abandon space activities unless a medium-sized project were to be undertaken by ESA every year. In fact, ESA's projects were then no bigger and no more ambitious than some national ones. The ESA program was at risk of becoming subcritical and of being unable to deliver what was requested from the scientific community and from industry.

That situation clearly demanded drastic action and, above all, an increase of ESA's Science Program fixed budget. Of course, it was not sufficient just to publish a shopping list of desired projects as support for the requested expansion. It was necessary to elaborate a coherent, well-structured plan, whose value could convince the politicians, and to give firm estimates of its costs to reassure the national financial authorities. This became Horizon 2000, through which ESA is now in a position to develop the facilities required by the community in Europe and to exercise the necessary coordination among the different national activities and plans, thereby avoiding unnecessary duplication and optimizing the European space science program.

National Space Organizations and ESA

In apparent contradiction to the above, more and more Member States feel inclined toward creating their own national space agencies. This proliferation can be considered as a positive sign of the growing importance attached to space activities in Europe, but the obvious danger exists that the most enterprising countries may generate centrifugal forces which may prove detrimental to the concept of a coordinated space effort.

Furthermore, the introduction of optional programs at ESA can be seen as a deviation from the original spirit of the Convention, which foresees the preeminence of a transnational spirit of cooperation, and has, to a certain degree, exacerbated this centrifugal tendency to promote programs more directly related to national interests. The proliferation of optional programs has, nevertheless, made ESA a very lively and enterprising organization, and, somewhat paradoxically, has reinforced the cohesion of its Member States.

The success in the implementation within the ESA program of the priorities of the more ambitious countries, which are also those which possessed strong national space organizations, has led the other countries to create their own. The case of France, though not unique, is very typical in this respect; clearly, the success of CNES as the arm of the French space policy has been an example that several countries have been keen to copy. It is well known, in fact, that the existence of CNES, whose creation and level of support reflect the importance attached to space activities by French politicians as far back as de Gaulle, has been a determining element in the shaping of both French and European space policy. This influence is most likely the reason behind the proliferation of space organizations—which should be a subject of great satisfaction for CNES.

Through a well-prepared and well-coordinated policy, with sustained financial support, France has been the main contributor to the ESA programs ever since the creation of the Agency. Having identified its own national priorities, it could afford, with some chance of success, to trigger the crisis which in the early seventies shook ESRO and ELDO and led to the creation of ESA. France forced the introduction of application programs, including meteorology—the very successful Meteosat program was French at the origin, prior to its ''Europeanization'' in ESA—the development of Ariane, and later, with partial success, of Hermes. CNES has extended its competence in nearly all the sectors of space activities, sharing its resources between a strongly national program— including bilateral cooperative ventures, in particular with the United States and the former USSR—and the European program.

Obviously, the coexistence in Europe of two very strong and powerful space organizations like CNES and ESA, pursuing nearly identical goals, could not but create some kind of competition or conflict at times. The French financial contribution to ESA comes from the budget allocation of CNES, and, not surprisingly, some in France have claimed that the money allocated to ESA would be better employed if it were spent for the benefit of national or bilateral programs controlled directly by CNES. The French delegation is, for the most part, made up of CNES engineers and administrators, usually well briefed and prepared and in the best position to play a dominant role in all the deliberations of the Council, the SPC, and the various delegate bodies.

Occasionally their dominant position makes the job of the chairman a difficult one. Back in the 1980s, for example, no one around the oval table of the Council will ever forget that long debate on a now forgotten issue, when the French delegation, led by Frédéric d'Allest, then Director General of CNES, was particularly reluctant to accept the unavoidable compromise. The chairman was Hubert Curien, who was also at that time the President of CNES, and, as such, its supreme authority. In order to lead the debate to a close and reach a momentary consensus, Curien addressed the delegations one by one as usual and, coming to the French, said pointedly: ''I am quite sure that the French Delegation can re-analyze its position. Monsieur d'Allest, can you accept the consensus. . . ?'' The compromise was reached without delay. Nevertheless, cooperation between ESA and CNES has been a key element in the success of European space activities overall. Certainly, as far as space science is concerned, throughout the 1980s and in the 1990s as well, France has been one of the strong and faithful supporters of the ESA program.

As far as the mandatory scientific activities are concerned, the funding organizations may not necessarily be under the authority of the respective Minister for Research. This is in particular the case in the United Kingdom, where the delegate from the British National Space Centre (BNSC) reports to the Department of Trade and Industry, while the funds for space science came, until recently, from the Science and Engineering Research Council (SERC), under the authority of the Minister for Science and Education. In the United Kingdom, space science is facing competition with nearly all the branches of fundamental science, such as ground-based astronomy for example. Hence, the degree of support for space science in the United Kingdom, and, in particular, the funding of the contribution to ESA's Science Program, is the result of decisions made by the Research Council itself, where priorities are established through the peer review system, not necessarily reflecting those which the BNSC might be willing to adopt. Such a peculiar situation finds solutions thanks to the high degree of diplomacy and of conviction of the UK delegate at the SPC and at the Council. On several occasions, however, the UK delegation has been particularly difficult, especially when the increase of the Science Program budget came under discussion. This was the case in 1988 when the continuation of the annual 5 percent increase necessary for the proper implementation of Horizon 2000 required a unanimous decision by the Council. The UK delegation conditioned its endorsement on the outcome of a general review of the management of the Science Program. This review was conducted successfully under the chairmanship of the German scientist Klaus Pinkau, and did not reveal any fault. Eventually, in December 1990, the United Kingdom voted, together with all other delegations, the continuation of the 5 percent increase until 1994, placing Horizon 2000 on a firm financial basis.

Figure 10 clearly illustrates the diversity of national organizations which contribute to the funding of ESA and whose representatives sit on the various committees and on the Council. This diversity may indeed cast some doubt on the efficiency of the overall European space effort. Like France, most Member States have devoted a large proportion of their activities to bilateral or multilateral cooperation with the United States, the USSR, Japan, and now China. For these countries, it may seem strange that the Europeans, while representing an overall GNP larger than that of the United States, are not able to contribute more substantially to the international space effort. Figure 11 and Table 6 compare the contributions of the main space partners in the world for 1990, illustrating the very peculiar situation of Europe. One might ask whether Europe would do better in space without ESA, with just an uncoordinated juxtaposition of national space programs. Or whether coordination might be secured without ESA. The success of ESA provides the answer to the first question. As to the second, if ESA did not exist, some other organization would have taken up its role. The present insistence of the European Community on taking a more leading role in the definition of European space policy is just a proof of this. Certainly, the remedy is to coordinate as well as possible the national and the common European activities and to plan the future together, so that each partner recognizes its interests in the overall effort and is fully willing to be part of the common enterprise.

ESA's industrial policy, which promotes national interests through an overall space program competitive with that of the United States, is an impor-

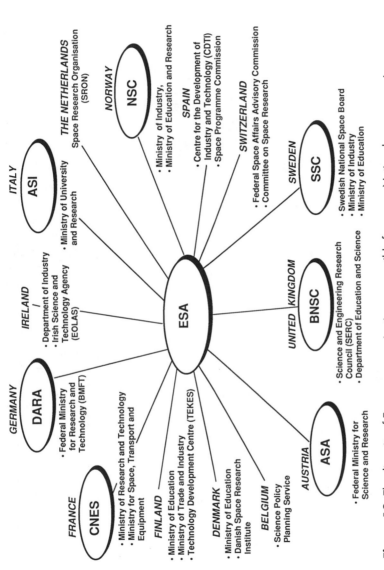

Figure 10. The diversity of European organizations responsible for space. National space agencies are identified by an oval.

1990 Civil Budgets for Space* Throughout the World (USD in millions)

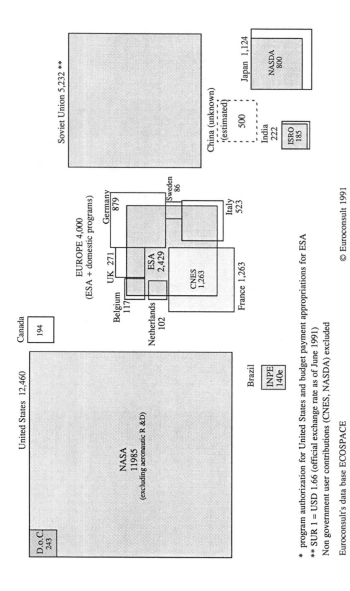

* program authorization for United States and budget payment appropriations for ESA
** SUR 1 = USD 1.66 (official exchange rate as of June 1991)
Non government user contributions (CNES, NASDA) excluded

Euroconsult's data base ECOSPACE   © Euroconsult 1991

Figure 11. Comparison, in millions of dollars, of the world's most significant space programs, 1990. The largest dark gray square at the center represents the ESA budget, to which the amounts spent nationally by the Member States must be added to obtain the overall European civilian space effort. (Source: Euroconsult)

tant factor in gaining support for this enterprise. This support is stronger from the smaller countries, whose national activities are too small to compete alone. But ESA offers more than just the security of some coordination. To justify its existence, the European space effort must add a new dimension to the purely national activities, a dimension which rests first on the expression of a political willingness to create an integrated Europe and to promote space activities as a determining element of European integration. In addition, the partners must identify what their particular interests are and what is their place in the overall program. Hence, it is necessary to promote optional programs together with a voluntary industrial policy, and to undertake missions of a scale well above what a single country could do. Therefore, it is necessary to define a Scientific Program based on the aspirations of the scientists in the various Member States. All three ingredients, political, scientific, and industrial, are vital to the recipe for successful cooperative effort.

▪ THE CAPRI MEETINGS

Horizon 2000 has had a profound effect on the organization of space science in Europe. Very soon after being accepted by the ministers in Rome in January 1985, it became a reference program and, as such, could play the long-needed coordinating role between the European space program and the various national activities. This perception became more concrete in the framework of the so-called Capri meetings.

In 1987 the Italian delegate to the SPC, Saverio Valente, who was Lord Mayor of the Island of Capri, invited ESA to hold meetings of this committee every second year near his villa, which had a marvelous view of Naples and Mount Vesuvius. The invitation was accepted wholeheartedly by all the delegations. The first meeting took place in May 1988, with Vittorio Manno in charge of its organization, and the next three in 1990, 1992, and 1994, with Giacomo Cavallo in charge. These meetings are now an accepted procedure, whereby the SPC delegates, the members of the Astronomy and Solar System Working Groups and of the SSAC, join forces to ''brainstorm'' about the national and ESA space programs, while saving some energy to take part in local folkloric dances and to enjoy the generous hospitality of their host, at the party which usually concludes their efforts at coordination, with the Bay of Naples in the background. Under such conditions, how could these efforts fail? And indeed they do not.

The meetings generally start with an up-to-date description of ESA's and the national activities, including technology research and recent developments. Matters of general interest are also addressed, such as the archiving of data, the setting up of a small satellite program, or the attitude to adopt toward international cooperation. The task of synthesizing the various inputs is left to the working groups and to the SSAC. Only space science, as covered in the ESA mandatory program, is considered in the framework of these meetings: infrastructure, telecommunication, Earth sciences, and microgravity programs are not discussed.

The meetings also offer an opportunity to compare the relative financial and technical efforts of the Member States, and to estimate the complementary level of effort to be undertaken nationally. Table 6 gives a breakdown of the various contributions as reported in May 1992. These data also allow a more precise comparison between the level of the European effort in space science and that of the United States.

In 1990 the overall effort in Europe added up to about 400 MAU. The equivalent activities of NASA for that year totaled $1,250 million, excluding the costs of the launchers—the Hubble Space Telescope and the Ulysses mission were launched at a cost of more than $300 million each. The ratio of NASA spending to European spending, including the launchers in both cases, is nearly five to one. These figures, and those relating to the overall respective spending of the United States and Europe, as shown in Figure 11, demonstrate that Europe overall is still lagging behind the United States. This comparison may explain why ESA often plays the role of a junior partner in its cooperative ventures with the United States. On the other hand, the Europeans spend approximately five times more than the Japanese: the budget of the Japanese Institute of Space and Astronautics Studies (ISAS) totaled $150 million in 1992. A quantitative comparison with the Russian effort is still difficult to make.

■ TOURING EUROPE

As we have seen, the relations between ESA and its Member States are very complex indeed: on the one hand, ESA must formulate a European program, based on the needs and the requests of its users, and in coherence with the expectations and the capabilities of European industry; on the other hand, it must be prepared to understand and accept national priorities giving rise to national programs and to coordinate them into an optimal overall European effort. The tools are those described earlier: on one side, the scientific and industrial committees and the long-term plan; on the other, presentations, such as those at the Capri meetings, of national scientific activities.

It is important to underline another tool, which in our view is of the greatest importance. This is a deliberate policy of regular and periodic visits to all the Member States, in order to discuss directly, with the delegations and the representatives of the scientific community and of industry, their respective problems and their wishes. These are not emergency meetings set up in order to address problems *a posteriori,* but preplanned events between ESA and the delegations. Through them, understanding increases on both sides, and hidden problems are brought to light and ways to cope with them identified. These meetings are very helpful for achieving the proper solutions for problems and for rebuilding the mutual confidence that is sometimes put to severe tests in the Council meetings.

It is difficult to give a complete survey of these visits, and of the type of problems that are dealt with or that arise during the meetings. They are very different, depending on whether they arise from the smaller or from the larger Member States, from partners in other agencies, from those which have or those which do not have national programs, from those which border the Mediterranean or those which enjoy the long days and bright nights of summer in the north. All are equally essential in the implementation of a European space policy. Indeed, there is material for quite a number of fine stories about these experiences, including lighter ones concerning the different styles of receptions organized by the various countries: from being sumptuously entertained to being left nearly alone; from being driven by a chauffeur in a Bentley to being urged into a shaking taxi to the airport, albeit with excellent bottles of wine in our hands; the meeting that became a confrontation between national official representatives and some of their scientists; being questioned about corrosion of airplanes from saline humidity by a general who did not know that ESA was a space agency, although he was responsible for it in his own national organization. Everywhere, however, we have been received with exquisite courtesy.

The process of coordination is fundamental to the relation between ESA and its Member States. It began only in 1988 for the space science programs and is certainly far from being completed in all domains. Even a cursory look at the presentations given at the Capri meetings makes clear the central role of Horizon 2000, and how much coordination and coherence have grown out of its existence. This program has definitely changed the situation in space science in Europe. Even though robust national programs do exist, also with considerable international involvement with NASA, Japan, and the former USSR, ensuring their coherence with Horizon 2000 is becoming an everyday part of the ground rules.

Notwithstanding all these positive elements, many of those present in Capri, confronted with the variety of methods, procedures, accounting systems, and currencies used in Europe, must have given a longing thought to the old Roman emperor who, from the same island, presided over a more homogeneous empire encompassing the greater part of ESA's countries.

4

INTERNATIONAL CONNECTIONS

In the area of space science, as we have said, ESA enters into cooperation with other agencies only in the domains which are its responsibility, namely the provision of general facilities: platforms, satellites, launchers, and operations in orbit. International cooperation concerning payloads carried on board the satellites falls under the responsibility of the experimenters themselves and of the Member States which fund them.

In all its cooperative ventures, ESA follows the principle that collaboration with other agencies should be of advantage to both parties. Although the discussions and bargaining were often conducted in a business-oriented atmosphere, with each of the participants trying to secure the greatest advantage, there was always a realization that an exaggerated advantage obtained in one particular situation would have negative effects on long-term relations and on the fate of future cooperative ventures. Also highly important to the Executive was the concept that cooperation, and at least a certain degree of coordination among the various agencies, not only would best suit the interests of science and of the scientists but would also prevent wasteful duplication of effort, especially at a time when more resources are necessary for the development of increasingly complex and ambitious missions.

International collaboration in ESA's Science Program evolved from an essentially exclusive relation with NASA to tentative openings toward the USSR, Japan, and other nations, and finally achieved a worldwide dimension within the Inter Agency Consultative Group for Space Science (IACG), to which we must add numerous cooperative agreements in the domains of application and manned missions. This evolution mirrored the changes occurring in Europe in the sophistication of projects, the size of the scientific community, and the rising maturity and self-assertiveness of the European space program. With the formulation of Horizon 2000, a new dimension was introduced, which allowed forward planning over two decades and the possibility of cooperation and coordination *a priori,* with an optimization of the overall effort and of the financial resources available for space research in the world. Table 7 lists the missions of ESRO and ESA conducted in cooperation with other organizations or agencies.

Table 7.   ESA missions involving other agencies

Mission	ESA share	Partner	Partner's share
ISEE-2	Spacecraft	USA	ISEE-1, ISEE-3
IUE	Solar Arrays, S/C operations (1 shift)	USA	Spacecraft, launch, and operations (2 shifts)
		SERC/UK	Science instruments and detectors
Giotto	Spacecraft, operations	USA	Deep Space Network
		USSR	Vega-1, Vega-2
		IACG	Coordination
HST	Solar arrays, Faint Object Camera, operations	USA	Spacecraft, instrumentation, and operations
Ulysses	Spacecraft, operations	USA	Radioisotope thermionic generator, launcher and upperstage, scientific instruments, DSN
ISO	Spacecraft, launcher, and operations	USA	Second ground station
		USA, Japan	Operations
Soho	Spacecraft	USA	Launcher, hardware, and scientific instruments; operations
		IACG	Coordination
Cluster	Spacecraft, launcher, and operations	USA	Scientific instruments and hardware
		UK, China, Hungary	Cluster Science Data System
Huygens	Titan probe, operations	USA	Launcher, Cassini spacecraft, scientific instruments, and DSN
XMM	Spacecraft, launcher, operations, telescope (Option 1)	USA	Payload
		Russia	Launcher
	Integral spacecraft, operations, launcher (Option 2)	USA	Operations

Note: Within the ESA system, payloads are paid for by the Member States.

▪ NASA

NASA has been by far, and since the late 1960s, the predominant non-European partner on ESA's scientific missions. This has made it necessary for ESA to have a permanent office close to NASA, in Washington, in order to follow not only the evolution of U.S. space policy in general but also the situation of its cooperative projects.

The pattern of cooperation with NASA changed over time. In the initial years of ESRO, cooperation took the form of gratuitous launch and launch services provided by the United States in exchange for some payload sharing on the spacecraft (ESRO-1A and 1B, ESRO-2B). The program evolved in the 1970s into a purely European one, based on the utilization of U.S. launchers purchased by the Agency (ESRO-4, TD-1, HEOS 1/2, GEOS 1/2, COS-B). That was the period of expansion of the ESRO and ESA program, in a context of increasing space budgets, at a time when the size of projects could still be maintained at a medium level and the costs kept within reasonable limits. This age came quickly to an end, with the size of projects continuously growing. Space missions gradually became more sophisticated and costly. Observatories, open to the entire astronomical community, emerged, as well as multi-component projects. A period of more intense and more complex cooperation with NASA then began. Meanwhile, space science had become a mature discipline, attracting an increasing number of groups and individual scientists, imposing higher demands on its projects. It was thus necessary for the Scientific Program, handicapped since 1972 by an inflexible funding ceiling, to seek international cooperation in order to make available to European scientists the advanced facilities they required. Thus, roughly 50 percent of ESA's program was conducted in cooperation with NASA, with missions like ISEE, IUE, ISPM, the Hubble Space Telescope, paralleling purely European missions such as Exosat, Giotto, Hipparcos, and ISO (with some small participation by Japan and NASA).

While, in general, the cooperation with NASA has been an essential element in the successful development of European space science, and extremely beneficial to Europe, it has also involved some difficulties because of the unequal weight of the two partners, and the quality of the relationship has varied from case to case. For example, in the cases of ISEE and IUE it was flawless, while in the case of ISPM, now renamed Ulysses (and described more extensively in the next chapter), it accumulated difficulties and revealed how little binding on NASA is the Memorandum of Understanding (MOU) stipulated between it and the other partner, and how vulnerable are the U.S. space projects to the process of annual budget approval.

(The MOU is the text that defines the respective responsibilities of the two participating agencies. It is understood that the ability of the two agencies to carry out their obligations is subject to their respective funding procedures. It is also understood that changes in the content of the mission of the MOU, or its scientific scope, will be agreed upon mutually by the managers of the two agencies. The MOU is founded on a "best efforts" basis. Once the MOU is signed, a project plan is established which describes in more detail, usually technical, how each agency is to discharge its responsibilities.)

A less celebrated, but ultimately more significant case from the point of view of Europe's identity, was that of the joint Temple 2/Halley cometary mission. Quite differently from ISPM, this project was only in the study phase. But there again, because of the unfavorable financial outlook, NASA decided to withdraw its part. Out of this situation, the purely European Giotto was born.

The Hubble Space Telescope (HST) was extremely challenging, both because of the ambitious scope of the mission and because of its cooperative aspects. Within Europe, prodded by a direct and, needless to say, forceful admonition by Giuseppe Occhialini, who was then advising the Executive— "What does ESA do while NASA is planning the Space Telescope? If you can't beat them, join them!''—ESA proposed to NASA to contribute three elements to the mission: the solar arrays, participation in the operations, and the development of one of the noble focal plane instruments, the Faint Object Camera (FOC). Some U.S. scientists would have preferred ESA to supply the telescope itself and wanted to retain control over the instruments. They expressed some doubts as to the technological capability of Europe to develop the FOC. The result was that NASA formed a "tiger team" which was sent to tour the space institutes and industrial firms of Europe. To the Americans' surprise—but not to ours—NASA was forced to admit that Europe could indeed do the job. In return for its contribution, ESA requested a guarantee of 15 percent of observing time on all instruments (in reality, European astronomers at present manage to obtain an average proportion of 20 percent after selection through the peer review system). The argument was heated on both sides of the Atlantic. In Europe, some questioned why ESA should invest so much in the program in order to have access to what the astronomers would get anyway through normal competition, since NASA intended to open its Announcements of Opportunity to the entire scientific community.

The overall agreement with NASA required an uncountable number of meetings, in particular with Nancy Roman, who was responsible for astronomy at NASA. The final details of the agreement were worked out by her for the NASA part and by Duccio Machetto and Vittorio Manno for the ESA part, during a pleasant dinner at her home. The agreement was celebrated with a bottle of wine that the host had purchased especially in honor of these two Europeans from a wine-producing country.

After Hubble and Ulysses were accepted in the ESA Science Program in 1977, they were plagued by delays which, while not of NASA's making, were nonetheless costly and frustrating. Hence, while by and large successful, cooperation with NASA has also had some negative aspects which have engendered basic questions. For example, how credible was cooperation when one of its basic instruments, the Memorandum of Understanding, had no legally binding effect on one of the partners? How could the different funding procedures of the two parties be reconciled to give the same guarantee of continued funding? What degree of dependence on international cooperation could ESA afford for its program? How much more resources should European Member States provide to allow ESA more independence in its scientific choices? It is thus not surprising that after the wave of projects in collaboration with NASA in the late 1970s, the SSAC and the SPC would subsequently approve more European-controlled projects such as Giotto, Hipparcos, and ISO.

What about cooperation between the two agencies in the context of Horizon 2000, a plan that was not conceived in isolation or in competition with other agencies' programs? In essence, Horizon 2000 is open to cooperation and to joint missions which are of mutual benefit and are respectful of the European priorities and aims. One of the four cornerstones of Horizon 2000, the Solar Terrestrial Science Program (STSP), made of the combination of the Cluster and Soho projects, is already in a state of advanced development in cooperation with NASA. The X-ray Multi-mirror Mission (XMM), the second cornerstone, will have a substantial U.S. involvement in the focal plane instrumentation, and both remaining cornerstones, the Far Infrared Space Telescope (FIRST) and Rosetta, although presently defined within a purely European context, could be part of a cooperative approach with NASA and, potentially, other space agencies. These cornerstones are leading objectives of European space science and, as such, should naturally be under European control, with international cooperation coming as an add-on on top of a mission which the Europeans should be able to conduct with their own means, and which should not depend on major decisions being made by other organizations.

Now, the full pattern of cooperation between ESA and NASA is easily discernible: a period of dominance by NASA in the infancy of ESRO, followed by a period of intense cooperation, with ESRO generally in the role of the junior partner, and then a period characterized by a strong European desire for more independence. This third stage was facilitated by the successful development and exploitation of the autonomous launcher Ariane. With Horizon

2000, ESA entered a period of genuine partnership, made possible by the maturity of its Science Program, by the advanced status of European technology, and by the very existence of an overall long-term plan. Thus, in STSP, the rules are inverted with respect to the Hubble Space Telescope, with ESA playing the major role and NASA the junior one.

In addition, the long-term approach proves to be a useful tool for ensuring complementarity and coordination between missions envisaged in the same scientific disciplines. Thus, XMM is complementary to NASA's AXAF and FIRST to NASA's submillimetric explorer, SMIMM. It does make sense that these somewhat parallel developments should be coordinated, with reciprocal access to the facilities being granted to the two communities. Through this approach, continuity in the development of a given discipline, and the number of opportunities offered, will increase for the benefit of the communities on both sides of the ocean.

Space science offers a very good benchmark test for cooperation in a broader area. Thus, ESA also cooperates with NASA in the other disciplines: life sciences, fluid physics and materials sciences, and, of course, Earth sciences. Two joint working groups in life science and in materials science and fluid physics meet regularly, and plan joint experiments/investigations to be conducted with the International Microgravity Laboratory on board the space shuttle. Cooperation is also well under way in Earth observation. As an example, the selection of experiments on board the ESA and NASA polar platforms has been coordinated successfully.

Spacelab

In addition to space science missions, ESA's first major cooperative venture with NASA has been Spacelab, the second being the space station which we discuss in the next chapter. Conducted in an optional framework, with ten Member States participating (Belgium, France, Germany, Denmark, Italy, the Netherlands, Spain, Switzerland, the United Kingdom, and later Austria), Spacelab signified the entrance of Europe into manned space activities. Although a resounding technological success for European industry, the cooperation with NASA was not without difficulties. For example, ESA would have preferred the cost to be less (the cost-at-completion reached 140 percent of the original evaluation) and the Europeans would have been happier if the conditions of utilization had been more favorable to them. Spacelab was launched for the first time in November 1983 by the shuttle (Spacelab-1), and this was the only flight jointly programmed by the two agencies. Subsequent flights did not offer a guaranteed shared access by the Europeans, although scientists could get access through normal competition or through separate agreements, as in the case of the two German flights D-1 and D-2. On the other hand, NASA also suffered in the preparation of the mission from cost overruns, which reached up to 169 percent of the original cost estimate. Some American scientists even complained that parts of the flight hardware, in particular the computers and the pointing system, were of marginal, if not obsolete, technology compared to what they used on board American missions.

Douglas R. Lord, then director of the Spacelab program at NASA, expressed this opinion: "No program is ever 100 percent successful, but if any international effort is to be called a success, Spacelab qualifies. As a result, Europe now has a manned space system capability and the U.S. has an excellent manned laboratory system to use with its Space Shuttle."[1] And Reimar Lüst, former Director General of ESA, represented the opinion of the Europeans on this joint ESA-NASA cooperative program when he wrote:

> International cooperation does indeed depend a lot on the actual balance of power, but the benefits of cooperation cannot always be explained solely in figures. Just as many European firms today spend a lot of money to buy themselves into joint ventures with American and Japanese high-tech companies, in order to get knowledge on new technologies transferred into their firms, so ESA had to pay the price of Spacelab to acquire the basics of manned spaceflight.[2]

It might be useful to recall that in 1973, when the Spacelab Memorandum of Understanding was signed, Europe did not have any expertise in this domain at all. Furthermore, while NASA had just succeeded in a series of spectacular and historic manned missions to the Moon, Europe did not even have its own launcher, the development of Europa having been discontinued the preceding year after several launch failures. Spacelab was part of the 1972 package deal which also included the development of Ariane. One might say that without Spacelab Ariane would have had a more difficult time coming into existence. Another view can be expressed, considering the fact that since the early 1960s NASA had quite extensively helped the European scientists by offering them numerous flight opportunities, free of charge: Spacelab was one way Europe could reciprocate NASA's generosity.

Today Europe has demonstrated its competence in developing Spacelab and can also capitalize on the success of Ariane. It is no longer in a subordinate position. The negotiations which led to the establishment of the Memorandum of Understanding and of the Intergovernmental Agreement, based on the principle of "equal partnership," which defined the joint European participation in the international space station and which were signed by the participating countries in September 1988, have reflected this change of situation and attitude, as we will see later.

The Reciprocity Agreement

Overall, the cooperation between ESA and NASA has indeed been successful. It has offered European scientists access to facilities and missions that their relatively modest means would have prevented them from developing by themselves. However, again quoting Reimar Lüst: "When resources abound and opportunities are plentiful, a cooperative attitude abounds. When resources shrink, altruism takes a back seat."[3]

In the 1970s and early 1980s, NASA's generosity came under fire from some American scientists who requested that Europe reciprocate NASA's policy of international cooperation, through which scientists from outside the United States were able to compete and get flight opportunities on NASA satellites. Not surprisingly, the viewpoint in Europe was that this policy, which had allowed an increasing number of European experiments to fly on American missions, was very satisfactory. However, no agreements existed which would have allowed American scientists to respond to European Announcements of Opportunity on ESA or national missions. The political pressure was directed toward ESA—although ESA was one among many players in the game—and some SPC delegates requested the Executive to take the responsibility for Europe to reciprocate that policy, which, by and large, had been implemented through bilateral agreements between NASA and the individual Member States.

This looked rather odd, as ESA's brief is to develop projects for European scientists, and not for scientists from non-ESA countries. Furthermore, ESA could not be held responsible for reciprocating each individual agreement that Member States had negotiated with NASA, independently of ESA. The Americans, however, became rather insistent on the issue. That was one problem with which Roger M. Bonnet was confronted when he joined ESA in 1983, and one that he had to solve urgently if he wished to avoid a radicalization of the already difficult relations between ESA and NASA in the post-ISPM era of crisis.

He formed a joint committee made of U.S. and European scientists, selected on the basis of their past experience in cooperating with one another, with Kees de Jager, Jean-Louis Steinberg, Kenneth Pounds, and Johannes Geiss on the European side and Norman Ness, Arden Albee, Thomas Donahue, Larry Peterson, and Andrea Dupree on the U.S. side. The meeting was co-chaired by Frank McDonald, then NASA's Chief Scientist, and Bonnet. The committee had a very simple brief: to formalize the so-called reciprocity principle. It met only once, at ESA Headquarters in the fall of 1983, and Bonnet reported the conclusions to the November meeting of the SPC.

During the discussions, it became obvious that the major differences between the two agencies were not properly perceived by the Americans. Could they in fact be criticized, remembering Henry Kissinger once asking whether Europe had a telephone number? In other words, did an entity called Europe really exist? A simple question illustrates this attitude. At several occasions during the meeting, the Americans raised their eyes to the ceiling and, paraphrasing Mr. Kissinger, asked Bonnet: "But please tell us: who is Europe?" They could not understand why ESA was hesitant about reciprocating a policy which benefited mostly the Member States and on which ESA was never consulted.

The committee, after a spirited discussion, concluded that ESA should indeed reciprocate and should offer access to U.S. scientists on its own spacecraft, while at the same time encouraging NASA to pursue its open cooperation policy. Interestingly enough, the first time the reciprocity principle could be implemented, when ESA opened its call for proposals for the ISO payload, NASA discouraged its instrument investigators from proposing. At that time, ISO was considered a severe competitor to NASA's SIRTF project, an American version of ISO—still to be started—and NASA was not keen to push a U.S. involvement in ISO which would have diluted the resources it intended to devote to its own infrared astronomy program. It did allow American scientists to make proposals in the category of mission scientists, a participation which implied only travel money for taking part in the meetings of the ISO Science Team (IST). Martin Harwit from Cornell University—now head of the Air and Space Museum in Washington—applied and was selected. Since then, he has taken full part in the activities of the IST. In spite of this delicate start, U.S. scientists now take advantage of the reciprocity agreement. They are involved in the XMM mission and will, most likely, funding permitting, also participate in the future missions of Horizon 2000.

▪ THE USSR, RUSSIA, AND CENTRAL EUROPEAN COUNTRIES

Until April 1990—apart from a legal instrument regulating exchange of scientific information—ESA did not have a formal cooperative agreement with the USSR. Nevertheless, scientific contacts had been numerous before that, and rewarding all along. They paved the way for the official agreement which was signed on 25 April 1990 between ESA and the government of the USSR. ESA's Space Science Department, like several other European scientific institutes, had, for a long time, developed scientific instrumentation on Soviet spacecraft—mostly in the domain of plasma physics. However, the working conditions were not ideal, partly owing to restrictions on technology transfer imposed by the Coordination Committee on Multilateral Export Control. Also, ESA technicians and scientists were not allowed to attend the launches of the satellites for which they had developed instruments, not even knowing when the launch would take place. On one occasion, they learned of the launch in the *International Herald Tribune!* Notwithstanding these difficulties, even in the most difficult period following the occupation of Afghanistan, the scientist-to-scientist contact never broke down, and some degree of scientific cooperation was always kept alive. In fact, in some respects the scientists in IKI, the Space Research Institute in Moscow where most of the space science missions are undertaken, were introducing *glasnost* and *perestroïka* long before these words were officially introduced by Mr. Gorbachev. Paradoxically, ESA had to keep a low profile, especially at the Council, where some delegations considered the relationships with the USSR as their exclusive privilege. A more formal cooperation would have been at that time exceedingly difficult because, within ESA, cooperation with non-Member States has to be approved unanimously by the Council. Today, in the present political situation, cooperation with Russia has been officially endorsed and even recommended by the ministers.

Certainly, nothing before was done in the shadows. A formal agreement on exchange of scientific information has been in existence between ESRO and the Soviet Academy since 1970. Formal discussions were held annually, alternately in Europe and in the USSR, during joint space science program review meetings. Progress was not necessarily spectacular, but relations were maintained. Information was exchanged and protocols agreed on, even though making photocopies was not an easy issue at that time. On one particular occasion, however, within the Inter Agency Consultative Group (IACG), set up with IKI, NASA, and ISAS to coordinate their respective missions to Halley's Comet, the relations with the USSR ascended to international recognition. As described a little later, cooperation then leaped several steps ahead and brought the four agencies involved to an unusual degree of openness. In the meantime, events have been overtaking each partner and cooperation with Russia—and the former communist countries of Central Europe—is now envisaged in a completely different context.

After the signature of the 1990 official agreement, five working groups were created which would meet once per year. They covered space physics, biology and medicine, microgravity, Earth observation, and manned space transportation. A new step was made in 1991 just before the breakdown of the USSR, when the SPC authorized the Executive to negotiate formal cooperation with the Soviets in their Mars-94 mission, through the provision of the central mass memory of the spacecraft. Other projects are also being discussed, such as the participation of Russia in ESA's new project Integral, a gamma-ray observatory, and in the possible realization of a solar probe.

Another change occurred when in February 1992 the Russian government created the Russian Space Agency (RKA), an initiative which had the merit of clarifying a very confusing situation. Before that, so many entities existed, such as Glavkosmos, Intercosmos, and NPO Energia (the main industrial partner), that it was extremely difficult to know with which one to deal. While the already existing working groups, in space physics, Earth observation, and life science, continued—ESA flew several highly successful instruments on the BION-10 spacecraft—it became necessary to establish new ones in order to implement the new policy of cooperation as defined by the ministers in Munich in November 1991. Thus, the two Director Generals, Jean-Marie Luton and Yuri Koptev, established three groups dealing with missions to the Soviet space station Mir, manned space infrastructure, and technology related to space planes, and on 12 October 1992 they signed a joint declaration outlining the need to pursue cooperation in the classical domains while assessing cooperation in the domain of in-orbit infrastructure and arranging for ESA astronauts to fly on board the Mir station.

These perspectives were fully endorsed by the ministers when they met again in November 1992 in Spain. Since then, contacts between the two Director Generals have taken place on a regular basis, while the engineers on both sides discuss potential cooperation in the new domains. Because of this completely new context, the agencies decided to negotiate a new framework agreement which would replace that of April 1990.

Concerning cooperation with the Central European countries which formerly belonged to the Soviet Intercosmos organization, several contacts have been made with ESA, in most cases through the respective national academies of science. The approach taken has been both open and cautious, ESA analyzing individual demands as they come, on a case-by-case basis.

Hungary signed an agreement with ESA in April 1991, which was implemented with the granting of co-investigator status on Cluster to the Research Institute for Particle and Nuclear Physics of Budapest. Official contacts have also been made with Poland, Czechoslovakia, Romania, and Bulgaria, which may lead to similar agreements. A possible way to integrate these countries into the ESA program, without yet negotiating their possible membership because they are not yet ripe for this decision, would be through the Prodex program. Prodex would allow them to contribute to ESA satellites payload elements which would be developed entirely in the home country, with the guarantee of a 90 percent return, and 10 percent overheads for the management of the program being paid to ESA in convertible currencies.

■ JAPAN

Cooperating with Japan presents challenges of a different nature. In accordance with their philosophy that a scientific mission should be tailored to an "in-house" development, the Japanese undertake projects which tend to be rather small, albeit sophisticated and very often at the forefront of space science in their domains, but managed with a style that is hard to reconcile with ESA procedures. On one side is Japan, with a national program to be carried out within strict financial limitations, while, apparently, unencumbered by industrial constraints. On the other is ESA, with its task of realizing a supranational program and projects beyond the capacity of any one single nation, itself encumbered by strict rules on industrial involvement.

ISAS, the Japanese Institute for Space Science,[4] holds faithfully to its rule of launching one small or medium-sized spacecraft per year, and would be hesitant to accept a cooperation scheme with another agency that might jeopardize this clockwork schedule. Japan is clearly interested in cooperating at the level of payloads, but this is unfortunately of little interest to ESA, since payload development is not an ESA responsibility. Hence, there has been no hardware cooperation between ESA and Japan as of now, although collaboration has developed with individual Member States, in particular with the United Kingdom. Only in two cases was ESA involved in helping European scientists to participate in the utilization of Ginga and Asca, two satellites for X-ray astronomy, and reciprocally, Japanese scientists to have access to Exosat, the complementary facility at ESA.

As was the case with the USSR, the momentum picked up considerably with the arrival of Halley's Comet and the setting up of the IACG. The Japanese were integrated into and chaired international teams, and brought their essential contribution to the success of this joint effort. The new activities currently assumed within the IACG in the area of solar-terrestrial relations, linked to an expansion of the Japanese space science program, bode well for a deepening of the joint work between ESA and Japan. It is now accepted that ISAS will participate in the operations of ISO along with NASA, with Japanese scientists involved in an extra third shift of observation, permitting a maximum efficiency in the recovery of data from the spacecraft, originally operated through two shifts only, and a single ESA station. Within this agreement, NASA would contribute the second ground station.

ESA and Japan are holding regular yearly meetings, alternately in Europe and in Japan, involving both NASDA and ISAS, during which areas of potential interest are discussed in space science, Earth observation—in particular the cross-calibration of data between ESA's ERS-1 (Figure 12) and the Japanese equivalent JERS-1—but also telecommunications, and of course the

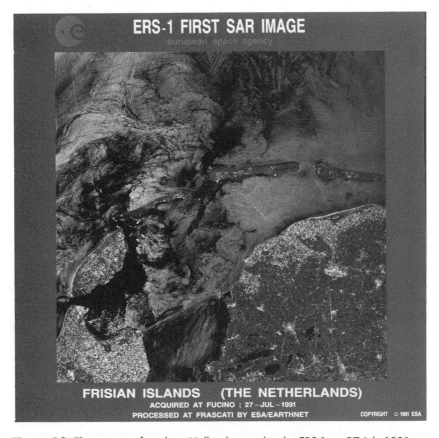

Figure 12. This picture of northern Holland was taken by ERS-1 on 27 July 1991, ten minutes before midnight at ebb tide. It covers an area of 100 × 100 km. The differences in the gray tone of the sea reflect different states of the surface: the whiter, the rougher the surface. Note the ship wakes extending more than 30 km. The resolution on the picture is 26 m, allowing a detailed view of the ground and in particular the fields, the urban agglomeration, and the aquatic structures.

space station, in which the Japanese fully participate. A new joint working group was established to assess possible cooperation in the area of space planes since both Japan (project Hope) and ESA (Hermes) have been studying such systems in parallel. ESA once envisaged opening an office in Tokyo, but the negotiations with Japan are at a standstill for the time being.

▪ OTHER COUNTRIES

India

India has at times expressed its desire to be associated one way or another with the space science program of ESA. This is quite understandable since there is a thriving scientific community in India and a very good school of optical astronomers and radioastronomers. In 1986 some were keen to participate in the projected very long baseline interferometry satellite Quasat, then under study at ESA. This participation never materialized, unfortunately, since Quasat was not selected by ESA. However, the Indian space program has been essentially oriented toward application satellites in telecommunication, and Earth observation, which is an area of national priority for India. In fact, this vast subcontinent, deprived of the basic ground infrastructure available to the United States or Europe, relies heavily on the space segment for the management of its territory for monitoring deforestation and replantation of waste land, for identifying water wells, and, last but not least, for establishing a network for instruction of its people in the remote parts of the country. Hence, the space segment is for India a critical and essential element of its future development.

Although matters did not proceed much beyond a framework agreement, including exchange of visitors and expressions of good will, ESA is impressed by the Indian space effort and by the seriousness and pragmatism with which it is being carried out. There is obviously a potential for future cooperation. It is therefore not surprising that the Indian Space Research Organization (ISRO) and ESA have negotiated an agreement permitting India to access the ERS-1 data with the ISRO ground station at Hyderabad.

Australia and Canada

In the space science domain, both countries expressed interest in some of ESA's projects. Australia in particular was keen to participate in the ultraviolet astronomy project, Lyman, under study at ESA in 1984–1985, and in Quasat. Joint studies were conducted, and areas of potential participation at subsystem level envisaged. In the process of selection, however, these projects were not retained and, unfortunately, new opportunities in space science have not yet been identified. For Australia, Earth observation is a privileged domain of cooperation, with a great interest in the ERS-1 data. An ERS-1 receiving station is installed in Australia. ESA for its part is interested in using the ground station at Perth.

Although Canada has a special agreement of close cooperation with ESA, it does not yet participate in its Science Program. It is mostly concerned with Earth observation, telecommunications, and manned space flights through its participation in the space station.

China

In spite of their strong space effort, in particular in the launcher area, the Chinese have had relatively few scientific satellites of their own, and contacts with ESA have not been very extended. Until recently at least, space science was not a priority for the Chinese. However, in 1991 the Center for Space Science and Applied Research of the Chinese Academy of Sciences at Beijing (CSSAR), replied to the ESA Announcement of Opportunity, mostly addressed to European institutes, to provide a data dissemination center for the utilization of the Cluster mission. That was clearly a surprise. The proposal was of high standard and therefore was retained. A formal agreement between ESA and the Chinese Academy has since been negotiated, opening a new avenue of cooperation. China has obviously a very high potential in space science and its scientists are keen to break the isolation which marked the period after the Tienanmen student revolution. They can soon become very powerful and very active partners. Like India, China is interested in getting access to Earth observation data collected by ESA's ERS-1 mission.

South America and Africa

The countries of Argentina, Chile, Brazil, Mexico, Colombia, and Ecuador in South and Central America, and of Kenya, Morocco, Tunisia, and Zimbabwe in Africa are mostly interested in Earth observation, telecommunications, and space technology, but have expressed no clear intention yet in cooperating with ESA in the area of space science.

■ THE INTER AGENCY CONSULTATIVE GROUP

With the approval of Giotto, ESA rejoined the club of space agencies which were developing spacecraft for observing Halley's Comet in 1986. Also in the club were ISAS from Japan and IKI from the USSR. NASA, while not having a dedicated spacecraft itself, was contributing to this unique international effort with the International Halley Watch, launching sounding rockets, and reprogramming on a new orbit the international spacecraft ISEE, renaming it on that occasion the International Cometary Explorer (ICE). All projects were independent of one another (Figure 13).

The concept that a global coordination of these independent initiatives would greatly benefit the overall scientific goal of cometary exploration took hold rapidly among the scientists and the agencies. In the course of 1981 it became apparent that each party could bring original contributions to the others, and that a joint effort would certainly yield a higher output than the sum of unrelated separate efforts. In addition, Halley's Comet was providing a challenging target by itself, a unique opportunity for worldwide scientific cooperation in space driven only by the thirst for discovery. The whole of the Halley's Comet epic would be dominated by this sentiment of a joint, historical, and unprecedented effort, coordinated among the four main space powers of the world. Never among the participants was there any argument about who would do the best job, who would stand up for the first prize. All agencies would be first! Their common effort would be first, and would remain in history for at least seventy-six years, to the time when Halley's Comet would once again come close to the Sun.

The four agreed, thus, to set up the Inter Agency Consultative Group (IACG), an informal body composed of agency representatives, with the task of coordinating all matters related to the observations of Halley's Comet from space. The first meeting of the group took place, at the invitation of the late Giuseppe Colombo, in 1981 in the city of Padua, where in 1303 the Florentine painter Giotto di Bondone painted a comet on the Adoration of the Magi fresco in the chapel of the Scrovegni—a comet that has been thought possibly to be Halley, although this is now denied by several historians of science.

The IACG immediately set up three working groups to deal jointly with such tasks as the modeling of the comet's atmosphere, the coordination of space plasma measurements, and the optimization of the spacecraft navigation. The International Halley Watch, set up to coordinate all ground observations, also participated in the work of the group. The IACG then met annually, rotating among partner countries, in an informal atmosphere which permitted

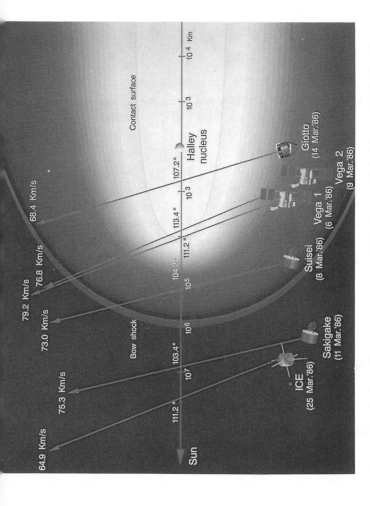

Figure 13. The fleet of space missions to Halley's Comet included Giotto (ESA), Vega-1 and Vega-2 (USSR), Sakigake and Suisei (Japan), and the International Cometary Explorer ICE (NASA). This international set of spacecraft was tightly coordinated through the Inter Agency Consultative Group (IACG). Thanks to this unique cooperative effort, Giotto could be directed to within 600 km of the nucleus, thereby providing the first detailed picture of a comet nucleus and very accurate results on the cometary dust and gas content at a very close distance to their source of emission.

personal contacts and generated friendship and trust, and brought up to a new level of collaboration the leading space agencies of the world.

The most decisive outcome of this joint activity was, without doubt, the "pathfinder" concept, diagramed in Figure 14, designed to allow the most precise targeting of Giotto with the comet nucleus. The nucleus could only be expected to be positioned by the International Halley Watch with an accuracy no better than about 1,000 km. This was more than adequate for the two Japanese spacecraft, Suisei and Sakigake, to accomplish their mission, and also for the two Soviet spacecraft, Vega-1 and Vega-2, which were to encounter Halley at distances of some 10,000 km and 8,000 km respectively, but this would not be precise enough for Giotto, targeted to encounter the nucleus at only 600 km. Fortunately, Giotto would be the last spacecraft to encounter the comet, and the cameras of Vega-1 and Vega-2, due to arrive a few days earlier, could provide Giotto with updated positions of the nucleus. These data would not be sufficient, however, unless they were combined with a very accurate positioning of the Vega spacecraft in space, to a precision of a few hundred meters. This accuracy, while not achievable by the USSR ground system, could nevertheless be achieved using NASA's Deep Space Network (DSN) to track the Soviet spacecraft. The combination of the data from the Vega cameras sent to ESOC (ESA's Space Operation Center responsible for the operations of Giotto) from Moscow, through an especially established fast line—still in place today and used for many other purposes—and those of the DSN could then yield a positional accuracy of the nucleus to some 40 km, compatible with the precision required for the final navigation of Giotto. Such was the "pathfinder" concept, whose success depended crucially on measurements done from the Soviet spacecraft and from NASA. Thanks to it, Giotto would be in a position to meet its objectives fully.

The operational strategy developed exactly as foreseen, with no flaws or failures in any of the numerous elements whose success was so determining, including the five launches, all perfectly successful and executed at the precise time. The International Halley Watch, under the direction of NASA, studied Halley from the ground, following and measuring its orbital parameters. Suisei and Sakigake passed the comet at distances of about 150,000 km and 7 million km on 8 and 11 March respectively, giving the first distant glimpses from space of this mysterious celestial body and of its large-scale characteristics. The two Soviet spacecraft took the relay, probing further inside the coma and zooming toward its mysterious power source to a distance of some 10,000 km. While Vega-1 was making the first ever close pictures of the coma on 6 March 1986, measuring its position with respect to the stars, NASA's DSN was tracking the Soviet spacecraft, measuring its position within the few hundred

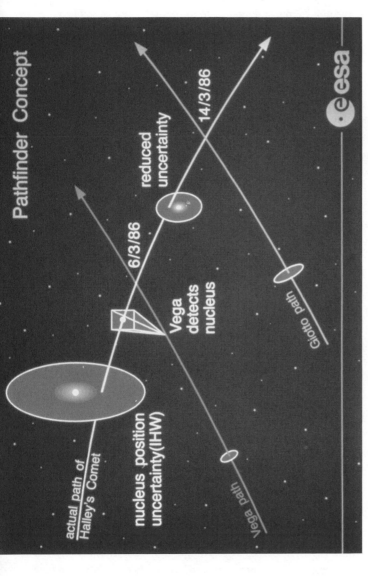

Figure 14. The Pathfinder Concept. The uncertainty with which the position of the nucleus of Halley's Comet was known from ground-based observations was ± 400 km. It is indicated by the large circle around the actual path of the comet nucleus. The cameras on the two Soviet Vega spacecraft, which arrived in the vicinity of the comet a few days earlier than Giotto, could locate the relative position of the nucleus. NASA, using its Deep Space Network, determined the absolute position of the two Vega spacecraft with high precision (the small circle around the Vega path), allowing Giotto to be targeted to encounter the nucleus with a greatly reduced uncertainty of ± 40 km.

meters' accuracy which, combined with Vega's picture, was needed to send Giotto to zoom on the still mysterious nucleus, with an accuracy of 40 km. All missions provided their expected and exciting results, among which the breathtaking pictures of the nucleus obtained by the Giotto camera were the most impatiently awaited by the scientists, the journalists, and the public.

This was the successful result of the efforts of all participants, of years of preparatory studies, of excitement but also, at times, of disappointment. It was the outcome of a joint and unique scientific endeavor, which cut across national boundaries and political and economic differences, at a time when the cold war was still regimenting the world, five years before the Soviet Union was to collapse. This first attempt at global cooperation, embodied in the IACG, succeeded fully because the goal was very precisely determined, because there was a maximum of informality and a minimum of bureaucracy, and because there was no exchange of hardware, all interfaces between the various participants being clearly established.

Nothing could have better symbolized this extraordinary peaceful dialogue than the solemn audience in the Vatican, on 7 November 1986 (Figure 15), during which the scientific delegations of all four agencies presented the results of this exceptional human endeavor to Pope John Paul II. In the richly decorated Sala Regia, the Pope greeted the space scientists as ''people of good will [who] seek to identify those areas of knowledge . . . which unite the human family rather than divide it . . . and thus . . . merit to be called peacemakers.''

With the epic study of Halley's Comet coming to an end, many in the four agencies thought that such an extraordinary experience should not fade away while Halley was on its way back to the depths of the Solar System. The IACG had brought together all four agencies over the period 1981–1986 as a team, regardless of political and other differences. In spite of some difficult moments, among which the most delicate was the invasion of Afghanistan, the scientists never stopped talking to one another and, more important, kept working for the future. During the meetings there were moments of tension indeed, but never any animosity. Very often, problems of a more political nature were solved, unconventionally, outside the doors. Ernst Trendelenburg, head of ESA's delegation until 1983, favored this approach. He would take his counterpart from the United States, the USSR, or Japan outside the meeting room, and the committee would wait until, a few minutes later, they would come back, all smiles, and with a solution. Roald Sagdeev, who was heading the Soviet delegation, once gave a masterly description of a laser beam device being developed in IKI for the forthcoming Phobos mission, and remarked that its power was a thousand times less than that foreseen within some projects

Figure 15. The ceremony at the Vatican on 7 November 1986, when the heads of the IACG delegations presented the results of the missions to Halley's Comet to Pope John Paul II.

of the Strategic Defense Initiative. That elicited some sour smiles from our American friends! But the IACG went forward to success.

So why let it vanish with Halley's Comet? Why not accept the challenge proposed by the Pope? It was decided not to abandon this winning team: the IACG would continue. That was not easy! The fundamental principle on which the IACG rested was that it should concern itself only with programs that were global in nature and in which all agencies had particular projects already approved. In other words, it should not become a planning organization, where bureaucratic considerations would supersede the scientific needs.

In the negotiations which took place in November 1986 in Padua, to define the "new" IACG, there was an all-European vision, an American vision, and a certain vacillation by the Japanese. Roger M. Bonnet was of the opinion that it was better to stop the joint activity if there was no genuine need to continue: it was better to stay with the stunning success of the Halley encounters than to tarnish it with some new more bureaucratically controlled phase. This potential danger could be felt when the Americans, led by Burt Edelson from NASA, requested that the name itself be changed into "the IACG for space science," probably in the fear that the IACG would take too much power. In spite of these difficulties, it finally got through, and the IACG is now in its second existence.

Its main objective is to coordinate the missions of the International Solar Terrestrial Science Program, again a worldwide cooperative effort in the exploration of the interplanetary medium and in the study of the relations between the Sun and the Earth (Figure 16). More than ten projects are spread among the four agencies in this domain, and the IACG is entrusted with the coordination of the respective science objectives, the optimization of the operational profiles, and the organization of the exchange of data. In the meantime, other scientific endeavors of a global nature have already appeared. In the area of very long baseline interferometry, the Japanese and Russian missions, VSOP and Radioastron, require the use of common frequency standards, together with the participation of the European network of radiotelescopes and the DSN from NASA. Similar developments are under way in the exploration of the Solar System, and in the domain of high-energy astrophysics, where some coordination would also enhance the expected scientific return of the missions planned in Europe, Japan, Russia, and the United States. Hence, the IACG continues its highly successful existence based on pragmatism, realism, and the search for an overall optimization of the space science resources around the world.

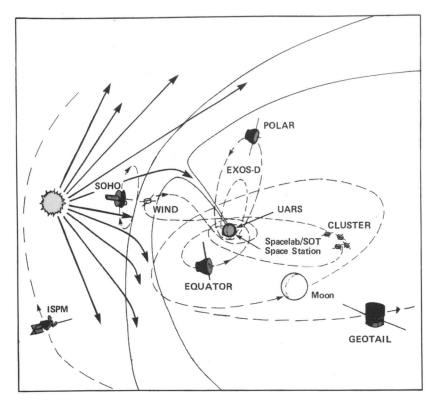

Figure 16. The Solar Terrestrial Science Program missions which are being coordinated by the IACG, involving in particular ESA's Soho and Cluster, ISAS's Geotail and Solar-A (Yohkoh), Russia's Interball, and NASA's Wind and Polar Satellites. In addition, several other missions are providing useful data on Sun-Earth relations.

▪ INTERNATIONAL COOPERATION IN THE FUTURE

The analysis of the evolution of the forms of cooperation of ESA with its major partners—particularly the United States and Russia—brings to light the different motivations which lay behind the word "cooperation," and allows us to discern the features and the quality of future cooperative ventures. For ESRO, some twenty-five years ago, cooperation was hardly a free choice; it was a scientific and technological necessity. At the end of the 1970s and in the 1980s the increased sophistication and costs of space facilities gave a new impulse to cooperation as a means to maintain a reasonable frequency of flight opportunities. Cooperation was no longer a condition *sine qua non* but a very efficient means of diversifying flight opportunities and being part of leading facilities participating in the very rapid advancement of space science. It became at the same time a means of sharing the costs of the development of large and complex systems.

Today space science evolves toward addressing issues that are multidisciplinary in nature and that require a global exchange of scientific data. Its all-encompassing role involves the world scientific community. Therefore, it is nurtured by and also nurtures international cooperation. Thus, the scope of planetary missions calls for a worldwide effort, and so do missions for the study of our local earthly environment, which rest on complex systems, involving several satellites and ground-based facilities as well.

Initially a way to learn and acquire experience, then a convenience, international cooperation in space science has now become a necessity. The question is no longer whether to cooperate, but rather when and how. But, to become more efficient, cooperation must be considered within a forward-looking perspective, with major projects clearly identified to encourage common planning. By comparing one another's programs, the space agencies can call on international cooperation at an early stage and make use of it, optimizing the worldwide scientific effort. At ESA, Horizon 2000, which spans twenty years, has indeed created that long-term forward-looking component, and has backed it with a budget stability that is secured through a level of resources established for five-year periods.

Cooperation is now an integral part of the space policy of the different agencies in the world. It gives them the opportunity to rationalize and optimize their planning and their resources, by coordinating the development of their related facilities, and by granting reciprocal access to their missions, thereby avoiding wasteful competition. It is a powerful instrument to enlarge the spectrum of possibilities offered to the scientific community, encouraging contacts and cross-fertilization between different disciplines or schools of thought.

However, along with these positive aspects, the danger exists that international cooperation may become a tool of power and interference, rather than a genuine partnership. Furthermore, bringing together partners with widely different procedures and schools of thought may upset the original intention and lead to inefficiency and frustration. Therefore, a certain code of ethics— a *galateo*—should be followed by the potential partners and the agencies in their international endeavors.

Such a *galateo* was proposed by Vittorio Manno during a round-table discussion at ESA in Paris in March 1988, on the occasion of a small international colloquium organized in honor of Roald Sagdeev (after his nomination as "Personnalité de l'Année," a distinction which is given to artists, politicians, and scientists). The *galateo,* as then defined, consisted of the following rules:

- International cooperation should be the tool for the development of an optimum worldwide program for the benefit of space science.
- Programs of the different agencies should be complementary to one another and not compete, while allowing a healthy overlapping.
- Agencies should exercise fairness, restraint, and respect for one another in the process of establishing and/or modifying their programs.
- Reciprocal access to one another's facilities should be the norm, and scientific exchange of data should be encouraged.
- Hardware cooperation should be based on mutual advantage and clear technical and management interfaces.
- Coordination among separate projects should be implemented whenever possible.
- Fair acknowledgment of the contribution of the other parties should be recognized and publicized.
- Upholding international agreements should have priority.

The participants happily endorsed this set of rules, which are well in line with those of ESA. Such rules are in the interest of everybody, the agencies and the worldwide scientific community, because international cooperation in space science is a necessity now and will be so in the future. As demonstrated by the IACG, cooperation on scientific objectives is feasible above political and ideological barriers and is one powerful way of fostering peace and education among the participating nations. Europe, encompassing many different cultures and civilizations, must show the way.

5

TWO CASES OF INTERNATIONAL COOPERATION

The Space Act of 1958 required NASA to conduct its programs within the framework of international cooperation. For NASA, cooperation was not only a policy established at the highest levels of the U.S. government, it was also a means to secure support for its missions in the Congress and the executive branch of the government. For ESA, cooperation offered more opportunities to European scientists and the possibility of participating in missions of a scope that they could not achieve alone. Hence, there was a shared interest in cooperating. However, the fundamentally different U.S. and ESA budgetary processes led to unexpected difficulties. The International Solar Polar Mission (ISPM) and, at several occasions in its turbulent development, the international space station illustrate these difficulties.

▪ THE INTERNATIONAL SOLAR POLAR MISSION

Few cases have been so revealing and telling about the inner workings of the U.S. administration and its relations to the Congress as the ISPM. This project, however, was born to be the paradigm of ideal cooperation between NASA and ESA. The concept of exploring the third dimension of the heliosphere with two spacecraft, symmetrical with respect to the plane of the Ecliptic, and hence capable of resolving the spatial and temporal features of the interplanetary medium and of the solar wind, originated very early in the 1960s in discussion between scientists in Europe and the United States. Later, the results obtained from Pioneer 6, 7, 8, and 9, showing that the heliosphere in the vicinity of the Earth orbit is highly structured, reenforced this concept, which quickly gained ground in the scientific community on both sides of the Atlantic. A sharing of responsibilities between Europe and the United States for a joint mission was agreed upon. It was based on quite independent hardware contributions developed on the two sides, and, of course, on the principle of no exchange of funds between the two parties.

Each agency was to develop its own spacecraft, with scientific instruments, whether of a European or of a U.S. origin, to be accommodated impartially on both, after competitive selection. Each spacecraft was to carry a core of instruments to provide fundamental data on various aspects of the Sun. ESA's spacecraft was spin stabilized while NASA's was three-axis stabilized, hence able to carry solar imaging instruments, with preference given to a white light coronograph and an X-ray/extreme ultraviolet telescope. The two spacecraft were to be launched in February 1983 by the shuttle, and subsequently inserted into an interplanetary orbit by an inertial upper stage. In order to gain sufficient energy and to change the plane of their orbit around the Sun, the twin satellites would first be sent away from the Sun, in order to intercept the orbit of Jupiter some seventeen months after launch. Jupiter's gravity field would be used to deflect the two spacecraft onto two symmetrical out-of-ecliptic orbits, which would bring them simultaneously to fly above the two opposite poles of the Sun about two and a half years later.

The first feasibility studies were carried out with the scientific support of a combined team of scientists sponsored by ESRO and NASA in 1974, and in 1975 the mission concept was reviewed at a symposium held at NASA Goddard Space Flight Center. At that time, within ESA, ISPM, then called the Out-of-Ecliptic Mission, was in competition with five other candidates to be selected in 1976 in the Science Program. Following the recommendation of the SSAC (at that time still called the SAC), ISPM was eventually selected in November 1977, together with the European participation in the Space Telescope. Among the grounds for selection it was stated that ''the dual mission, to which ESA with its spacecraft would make a major contribution, offers the basis for a clean interface and fruitful cooperation with NASA.'' In February 1978 NASA and ESA jointly selected the scientific instruments to be carried on the two spacecraft, and in March 1979 the two agencies signed the Memorandum of Understanding, which defined the distribution of responsibilities and tasks between them. Overall, more than two hundred scientists from sixty-five universities and research centers in thirteen countries were selected. On the U.S. side, the final go-ahead was given by the Congress, with the inclusion of ISPM in the fiscal year 1979 budget.

In order to clarify the different philosophies of the two agencies toward approval of projects, we must describe the approval procedure for science projects within ESA, as it sheds light on the incredulous reaction of Europe to the events which bedeviled the U.S. contribution to the mission. From an institutional point of view, a project approved for development within ESA's Science Program (whose budget is secured with five-year levels of resources voted by the Council) is approved with its financial allocations, defined within a certain cost-at-completion (CaC). Provided the cost for the development of the satellite, the launch, and the operations does not exceed 20 percent of the value estimated at the start of Phase B, the project will not be put into question by the SPC. This 20 percent level is, strictly speaking, not a legal constraint, because the SPC has authority over its total budget and can in principle allocate more funds to a project whose CaC escalates. However, the spirit of the rule is that a project in good financial shape should be able to be completed without threat of cancellation.

This procedure produces a strong incentive in industry, in the ESA management teams, and among the scientists to establish the CaC at a high level of confidence (within 20 percent) before approval, and to keep the lid on costs throughout the whole project. Statistics show that the CaCs of ESA's science projects are generally maintained within 20 percent of their approval value (see Table 4), unless the projects encounter unforeseeable technical problems, or are hampered by outside circumstances, design changes, or launch delays, such as those which followed the Challenger accident. A non-negligible advantage of the procedure is the very strong stability it gives a project once approved within the Science Program; in fact, in the whole history of ESA, no science project has been canceled. This is not to say that redirections and reductions in scope are not at times necessary in order to remain within the established CaC. But even in these cases, the projects all have been carried to completion. In one case—COS-B—the United Kingdom withdrew the funding of its experimental package and suggested that the project as a whole should be canceled. After an in-depth scientific discussion in a specially convened symposium, the United Kingdom withdrew its request, although the funds for the experimental package were not reinstated. The mission was salvaged by the intervention of ESA's Space Science Department, which became responsible for the missing scientific package, and the mission, launched in 1975, performed successfully and fulfilled its scientific objectives.

In the case of a cooperative mission with NASA, the Memorandum of Understanding (MOU) must be approved by the SPC and the AFC prior to being presented to the Council for a vote, which must be unanimous. The applicability of an MOU is always subject to the availability of funds for both parties, a condition which does apply with near certainty to ESA's Science

Program with its budget stability of at least five years. Only after Council approval is the Director General empowered to sign the MOU, which signature, unanimously agreed by the Council, commits the Member States to fulfilling their obligations as stated in the MOU.

The approval of ISPM followed this established procedure exactly and, strengthened by what the Europeans considered a legally binding MOU, the project appeared secure to the Europeans when its development started in 1979. Nobody in Europe doubted, at that time, that the United States would have the same reading of the situation as the Europeans. Unfortunately, the events which followed shattered this quiet conviction and initiated a new era in the relations between ESA and NASA.

The unexpected happened. The *New York Times* reported it on 9 May 1981 in a front-page article entitled: "US Dismays Allies by Slashing Funds for Joint Science Projects":

> It was a determined and angry Vittorio Manno who stepped into the chilly Paris morning and headed for Orly Airport on the Monday after President Reagan announced his budget revisions in February. He boarded a Concorde for New York, where he was met by two Americans—high officials of the National Aeronautics and Space Administration—in a private Air France lounge at Kennedy International Airport. The message Dr. Manno brought from Paris, where he is deputy science director of the 11-nation European Space Agency, was filled with injured indignation.

What had happened? In a short telephone conversation, a few hours before the public announcement by NASA, the acting NASA Administrator, Alan Lovelace, had informed the ESA Director General, Erik Quitsgaard, that it was NASA's decision to cancel its own spacecraft, thereby reneging on one of the provisions of the MOU. True, NASA still intended to honor the other provisions, but the crucial two-spacecraft mission would be amputated of one of its two components. NASA took pains to reassure ESA that the decision was purely due to new financial constraints and not to a scientific reevaluation of the mission, and that the financial provision of the MOU—"subject to the respective funding procedures"—was precisely what could not be met. Also, the proviso in the MOU stating that "any changes in the scientific scope of the selected experiments or of the final experiment complement will be established by mutual agreement between the NASA Administrator for Space Science and the ESA Director of Scientific Programs," would presumably have been satisfied by the advance information given by NASA to ESA's Director General. However, this last point particularly offended the Europeans, who could not accept that advance information on a decision that was already made and to be announced within a few hours could be construed as "mutual agreement." But let's see in detail the events which preceded this announcement.

In early 1980, still under the Carter administration, when the fiscal year 1981 budget was being prepared, NASA, which had been directed to make overall budget reductions in an election-year attempt to bring down the federal budget deficit, introduced a considerable cut in the ISPM request, which resulted in a two-year delay of the launch, from 1983 to 1985. Such a drastic change in the program had been discussed, and agreed upon, by NASA and, although with some irritation, by ESA officials. Within NASA, the shuttle development budget had to be protected by all means, and this led to important reductions in other branches of activities, among which space science was a designated victim.

However, the cause for the ultimate and unilateral cancellation of the U.S. spacecraft resided in fundamental changes brought about in the budget process by the election in November 1980 of Ronald Reagan as President, and, less visible to world eyes but very important for the NASA budget, by the appointment of David Stockman as Director of the Office of Management and Budget (OMB). Stockman introduced a top-down approach which effectively led the White House to control budget initiatives, a power which had previously rested with the various agencies and the committees of Congress. Together with this new budget philosophy, Stockman introduced a very strict policy of secrecy which made it very difficult for the federal agencies, including NASA, to open up their deliberations.

The original "blueprint" of the Carter administration for NASA's 1982 budget still included ISPM—albeit with a delay of two years—but the budget suffered major reductions in the revision by the OMB. The amended budget submission for space science was eventually close to 30 percent lower than originally foreseen. This led NASA to eliminate the ISPM spacecraft from its budget submission, without consultation with its partners and certainly not with their "agreement." ESA and its Member States were thus taken totally by surprise by NASA's announcement of the cancellation of its spacecraft. The outrage and incredulity in Europe were great. Outrage at the way the cancellation had been carried out, and incredulity that an international agreement would be canceled at all. This reflected ESA's stunned realization of the fundamental difference in attitude between the two organizations about the sanctity of a Memorandum of Understanding. In Europe, for the reasons described above, the MOU was considered as legally binding on its Member States, while it became painfully clear that this was not the case for the U.S. administration. That fundamental difference was to cast its shadow on all present and future international agreements between the two organizations. This realization, together with the lack of consultation with ESA, concerned the Europeans as much as did the scientific and financial ramifications of the elimination of NASA's spacecraft.

From the scientific viewpoint, apart from the serious degradation of the scientific objectives due to the impossibility of performing stereoscopic and imaging observations without the NASA spacecraft, about half of the instruments to be flown on the mission would not be used, and about eighty U.S. and European investigators were immediately eliminated from the project.

From the financial point of view, the funds already spent by the European scientists involved in the payload on the U.S. spacecraft—estimated at that time at about $15 million—would be irretrievably lost. Moreover, ESA had already committed the equivalent of $100 million to its part of the project. On the basis of industrial considerations of cost versus schedule, when the first launch delay to 1985 was introduced, ESA decided to adopt a "build and store" philosophy rather than stretch the development schedule. This meant that the development would continue unabated in spite of the announced launch delay. Hence, when it later came to the cancellation of the U.S. component, ESA had already committed about half of its funds, which would therefore have been lost if it had decided in its turn to cancel its own spacecraft, thereby killing the mission altogether. These considerations weighed heavily in the debate which ensued in Europe on what position to adopt with respect to the amputated project.

The ESA Director General, Erik Quitsgaard, and the Director of the Science Program, Ernst Trendelenburg, rapidly developed a forceful strategy, which was to be presented in a strong message to NASA. The delicate and somewhat unpleasant task of presenting the message to NASA was entrusted to Vittorio Manno, who represented ESA's Executive at the meeting with NASA on 20 February 1981, the one that was reported on the front page of the *New York Times* as quoted earlier.

ESA's position was clearly spelled out: the cancellation was a unilateral breach of the MOU, hence totally unacceptable, and ESA requested full restoration of the NASA contribution in accordance with the provision of the MOU. NASA responded that the cancellation was the result of severe budget cuts imposed on its Office of Space Sciences and Applications, but that it intended to continue supporting the other elements of its contribution: the launch, the provision of a radioisotope thermoionic generator (RTG),[1] its part of the payload, and the retrieval and dissemination of the data. NASA also expressed its hope that ESA would not in turn cancel its mission. ESA contested NASA's claim that it had met the MOU proviso that "both Agencies carry out their obligations subject to their respective funding procedures," arguing that by omitting its spacecraft from its budget request NASA had not even started the procedure.

The meeting in New York led to no agreement and ESA's Management Board decided on 24 February to take immediate action at the political level through the embassies of the Member States. A period of frenetic activity ensued, with a multitude of messages, telexes, meetings, discussions among ESA and representatives of its Member States. Within just twelve days an Aide Mémoire was approved by all the Member States and delivered to the U.S. State Department by three European ambassadors on 3 March. While the Aide Mémoire had a certain resonance in the State Department, where a committee was set up to study the problem, and elicited a supportive letter from Secretary of State Alexander Haig to David Stockman, Stockman's reply was uncompromising on the financial aspect, though suggesting that NASA might reprogram funds from within its existing resources. However, NASA's detailed budget for 1982, sent to Congress on 10 March, included no provision for the U.S. spacecraft.

The discussions between the ESA and NASA representatives were far from easy, the latter being torn between their loyalty to their management and their honest wish to salvage the cooperation with Europe. On one occasion, faced with the hard position taken by NASA, the ESA delegation solemnly left the room. No progress could be made. NASA's officials were sarcastically commenting that, while with expendable launchers a satellite once launched could not be canceled, with the advent of the shuttle even this could not be guaranteed anymore. The Europeans were smiling bitterly.

Having failed to alter the position of the U.S. administration, ESA shifted its efforts to Congress and its various appropriations and authorization committees and subcommittees then involved in the review of the budget proposal.

A Second European Spacecraft?

In parallel to this political activity, which saw ESA's officials—but never any NASA official—crisscrossing the skies between the two continents, ESA was elaborating on a technical proposal which could salvage the two-spacecraft configuration, though at a small additional cost to the United States. The main idea was that Dornier, the German prime contractor of the ESA spacecraft, would produce a second unit of the satellite that NASA would purchase, at a cost of—only—$40 million, representing a substantial saving for the American taxpayer who was supposed to contribute some $100 million to the cost of the original U.S. spacecraft. Indeed, this spacecraft would have been less sophisticated than the original U.S. concept. It had no de-spun platform and therefore could not carry imaging instruments, but it offered the great advantage of nearing the original capability of the mission at a minimal cost increase. In parallel, U.S. industry, through a strong lobbying effort, tried to save the U.S. spacecraft but did not succeed. The Executive was well aware that it was entering a minefield where these U.S. industrial considerations would play against its compromise offer, but that was the only possible financially acceptable alternative. To dispel fears of escalating costs, ESA committed itself to a fixed price with Dornier (to be updated for inflation) and assumed the risks of potential cost overruns. Throughout March and April, the ESA Executive repeatedly visited the Congress, the State Department, the Office of Science and Technology Policy (OSTP), and the OMB. Consideration was also given to contacting the National Security Council. It even came to the point where ESA itself was the source of information for NASA on what position the other U.S. actors would take.

The position of Congress was very sympathetic but unfortunately did not lead to a commitment. Congress would act positively provided the executive branch itself were to propose it. NASA, in late April, eventually accepted the principle of purchasing the second European spacecraft because of the scientific importance of having a second satellite. Hence, it notified the OMB that additional funds, albeit in smaller quantity than in the U.S. option, would be needed. Unfortunately, in spite of all its efforts and of the many encouraging signals ESA received, the situation was not developing. It was further complicated by the fact that a new NASA Administrator (James Beggs) was to be nominated shortly, and many found it correct or convenient to wait until he would take office in June.

Some positive signs did appear when the Congress added some money for ISPM within the 1982 NASA budget, sufficient in principle to allow NASA to maintain the two-spacecraft option. NASA was unwilling to proceed, however, unless additional money was granted by the OMB. It also argued that reprogramming resources would have threatened one of two other main projects both involving substantial international cooperation: Galileo (with West Germany), considered a key element in NASA's planetary strategy, and the Hubble Space Telescope (with ESA), strongly supported by the entire astrophysics community.

It became clear, however, that NASA never intended to initiate an in-depth study of the European second-spacecraft alternative. On 4 September, James

Beggs informed Erik Quitsgaard that NASA would not include any request for a second ISPM spacecraft in its 1983 budget submission then in preparation, arguing that a degraded spacecraft would substantially reduce the scope of the scientific results of the original two-spacecraft mission. A scientific argument (but leading to opposite conclusions) was thus introduced by NASA to justify nullifying ESA's efforts to restore ISPM as closely as possible to its initial configuration, from which it had departed because of NASA's financial constraints.

Having exhausted all possible means to reverse a deeply anchored NASA decision, ESA, in 1982, decided to proceed with its one-spacecraft mission. Meanwhile, the small spinning machine had been renamed Ulysses, and was scheduled to be launched in May 1986. The Challenger accident added more than 4.5 years to an already long delay, but Ulysses was successfully launched on 6 October 1990 by the shuttle Discovery. It flew by Jupiter on 8 February 1992, then bounced southward of the Ecliptic, contributing new insight into the Jovian environment. It is now boldly advancing on its interplanetary path (Figure 17).

Consequences

No one can deny that the ISPM crisis had a profound and lasting effect on the attitude of ESA toward NASA and on international cooperation in general. The MOU, which defined the respective responsibilities of the partners and governed their international commitment to the mission, obviously was interpreted differently by the two parties. To the binding character it had for ESA contrasted the sort of—loose—gentlemen's agreement it represented for NASA. History has in fact shown that, whenever it is faced with internal budget difficulties, the U.S. administration does not consider the MOU to be of any relevance to its internal deliberations. Because of the annual budget approval procedure in the United States, there can be no guarantee that a project may not be canceled at any time. In the United States, projects are subject to annual budget reviews and cuts are likely to occur as a result of technical or overall financial difficulties. That has been the case with ISPM, a scientifically sound project fallen victim to an overall budget cutback, and the first severe blow to cooperation in space between Europe and the United States.

In spite of the legal difficulties, the pressure to join hands again, to cooperate in space, slowly overtook the frustration of the moment. However, the spirit had changed. Europe would no longer accept being considered a subordinate participant. It had grown in maturity and self-assurance. Discussions in future

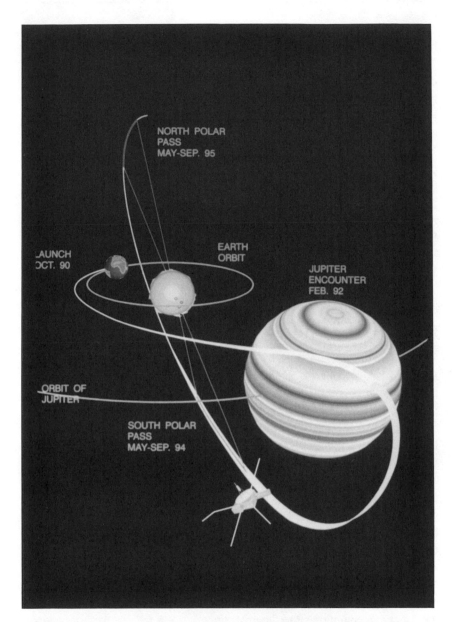

Figure 17. The present scenario of Ulysses, the first mission to explore the high heliocentric latitudes above the ecliptic plane and to fly above the poles of the Sun, involves only one satellite built by ESA.

cooperative ventures would be deeper and less one-sided. Consultations would be held in earnest and not limited purely to information.

Europe had lost a battle but had gained in resolve to go all the way on its own. Giotto came of age in 1980, but at the time of deciding whether or not the project would be conducted in cooperation, with NASA providing the launcher and part of the scientific instruments, the SPC decided that Giotto should be a purely European mission, launched by Ariane and without any U.S. principal investigator on board. When, in 1983, NASA exerted pressure on ESA to redefine the very sensitive all-European Infrared Space Observatory (ISO), and to merge it with NASA's equivalent project SIRTF (now postponed indefinitely), on conditions which were not acceptable to Europe, ESA stood its ground. The "ISPM crisis" therefore was not useless: it led to a new European attitude. The lesson had been learned. However, the intrinsic problems remained, shrouding the future of long-term cooperation with the United States in uncertainty and insecurity.

▪ THE INTERNATIONAL SPACE STATION

The history (at least up to now) of the biggest cooperative project between ESA and NASA, and the biggest international program in space ever undertaken, the international space station, is another illustration of the same problem. In 1984, following an invitation from the President of the United States for "friends and allies" to join the space station program, the administrator of NASA, James Beggs, initiated negotiations with Canada, Japan, and Europe. Having "neglected" to do so during the ISPM crisis, this time he crossed the Atlantic, and visited the various European capitals and the Seat of ESA, where he was received by his counterpart of the ISPM crisis, Erik Quitsgaard, and by the Council, at that time chaired by Hubert Curien. In spite of their somewhat traumatic recent experience, the Europeans were quite receptive to the invitation, but also keen to obtain assurance from NASA that the troubles they had just been through would, this time, be avoided. Hence, NASA found a different kind of European partner, determined to adopt a hard negotiating line and striving to obtain as much assurance as possible regarding the American commitment to the program.

The Europeans were looking for an agreement with the U.S. government which would include secured funding for the space station on a multi-year basis. The only secure way would be to establish a treaty between ESA, on behalf of its Member States, and the United States. A treaty would require ratification by the Congress, however, a lengthy process, with no assurance whatsoever that it might eventually be ratified. The idea was therefore abandoned by the U.S. negotiators themselves. It was eventually decided that an Intergovernmental Agreement (IGA) would be the preferred solution. The IGA would define the political and legal framework of the cooperative project, and its establishment would naturally involve the State Department. A Memorandum of Understanding would, in addition, be necessary in order to define the technical and the managerial interfaces which were being negotiated between NASA and the other space agencies involved: ESA, the Canadian Space Agency, and Japan's NASDA. For the sake of reinforcing the U.S. commitment to the program, it was envisaged to open this MOU also to the signature of the State Department, plus the OSTP and even the OMB, in order to make, as far as possible, the U.S. administration as a whole part of the "understanding."

During the early phases of the negotiations, a difference of interpretation—again—as to the meaning of these two documents appeared between the two parties. ESA, with its ISPM experience in mind, had the tendency to give more importance to the IGA, a document untouched as yet by the unsatisfactory experience of the ISPM MOU. The U.S. negotiators considered the IGA more or less at the same level as the MOU, an additional inconvenience which would also have to be negotiated and signed, and in no way assurance of multi-year funding. Indeed, no such assurance could be given, as the events which have marked the history of the space station year after year have evidenced. Nevertheless, ESA insisted on having an IGA. Faced with this necessity, the United States tried to negotiate one IGA individually with each of its partners, but had to accept, at their insistence, that only one IGA would be agreed upon by all of them. The IGA was then accepted as the principal document with which all the others—in particular the MOUs—had to be coherent.

The participation of Europe in the station was financed through an optional program named Columbus, with a leading participation of Germany amounting to 38 percent. It consisted initially of three main elements (Figure 18):

1. A Man-Tended Free Flyer (MTFF), to be serviced either by Hermes, the European space plane, or by the U.S. shuttle.
2. A pressurized module attached to the main body of the station, the Attached Pressurized Module (APM).
3. A polar platform dedicated to the study of the Earth and of its environment, and linked to the overall Earth Observing System of NASA.

Of these three elements, the MTFF was the most sensitive for the Americans. It was the blunt symbol of that European desire for autonomy, so strongly emphasized at the ministerial Council meeting in The Hague in November

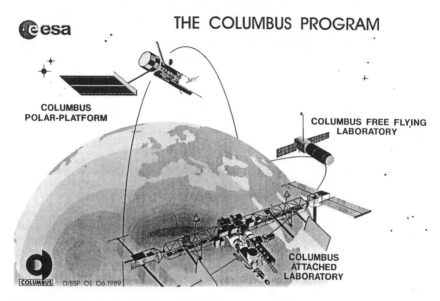

Figure 18. The originally planned European contribution to the international space station included three main elements representing ESA's Columbus program: (1) a pressurized module (APM) attached to the principal structure of the station, dedicated to materials and life science experiments; (2) a Man-Tended Free-Flying module (MTFF), accessible by both the U.S. space shuttle and the European Hermes space plane; (3) a polar platform for Earth observations. In 1992, the ESA ministers decided to abandon the MTFF.

1987—a desire which was going to be severely challenged at the two following ministerial meetings, in Munich and Granada, in 1991 and 1992. Moreover, NASA considered the MTFF an embarrassing complication to the station, and an element opposed to its own interest in the potential commercial exploitation of microgravity. However, ESA stood firm and refused to abandon this important element, which was paving the way toward autonomous manned operations, an expertise which the Europeans were still lacking. NASA yielded and accepted the MTFF as part of ESA's contribution. The APM was not originally part of ESA's proposal to NASA, but it was included during the later stages of the negotiation in return for NASA's acceptance of the inclusion of the MTFF.

Other difficulties appeared at the time of deciding what would be the management responsibilities of the various partners, NASA claiming that it should have complete control over all the elements, including the entirely European MTFF. ESA refused, reinforced as it was by the strong determination shown by all its ministers in The Hague—with the exception of the United Kingdom which declared itself not interested in this element of Columbus. To the surprise and dismay of their partners, the European negotiators showed a remarkable determination to analyze the Americans' requests thoroughly and to challenge their pressure.[2]

Between June 1985, when the MOU for Phase B of the station elements entered into force, and September 1988, when the IGA between all the partners was signed, the discussions focused on the legal as well as on the commercial and even the military aspects of the utilization of the station. The leading American negotiators besides NASA were the U.S. Office of Technical Assessment and the Department of Defense. The latter, in the summer of 1986, at the start of Phase C/D, caused strong perturbation when it expressed its intention to use the space station for military purposes, without specifying what these might be. The Department of Defense involvement, not foreseen at the outset, was particularly difficult for ESA to accept since several of its Member States—Austria, Sweden, and Switzerland—are neutral countries. In the midst of this difficult period, other problems arose. NASA, hit by continuous budget cuts, by the reality of the shuttle's greatly reduced annual flight rate, and by the fear of a repetition of the Challenger accident, undertook a substantial restructuring of the architecture of the station. This redesign exercise rendered the relations with its international partners even worse. While they were informed of the restructuring effort, they were not involved in the decision-making process. The rule of the *fait accompli* was, again, the basic management practice. Even the name of the station, Freedom, was decided unilaterally, by the President himself, without consulting the international partners, whose only ''freedom'' was to accept the decision of the leader.

Finally, to the great relief of all parties, a compromise was reached: ESA would keep its management responsibility over the MTFF and the Polar Platform, and NASA would remain responsible for decisions concerning the APM operations. In addition, barter arrangements on the utilization of the various elements were also agreed upon, through which ESA would retain 100 percent of its rights on the MTFF—NASA being offered 25 percent of its capacity—and 51 percent on the APM, the use of the Polar Platform and the definition of its payload being mutually agreed. Agreed also was the principle that the station would be used ''for peaceful purposes only,'' a term ambiguous enough to be interpreted according to the wishes of all the parties, all research activities conducted on a given module having to be approved by the party to which it belonged. This delicate period ended with the signing of the IGA in Washington on 29 September 1988.

The sense of equal partnership, and the determined attitude which the Europeans had shown in the course of the negotiations, were noticeably different this time. NASA now had to cope with the new European attitude. Then, gradually, NASA also adopted this spirit of partnership, and the relations between the two partners started to improve; at the working level, between the engineers on both sides of the ocean, relations have always been of the best cooperative quality. NASA had probably realized that the involvement of international partners in the station was an asset which reinforced the stability of that program and which could be useful to better resist the yearly financial battles with the U.S. funding authorities, in particular with the Congress, where in times of budget rigor and deficit reduction, more and more attention was being given to this visible and expensive part of the American space program, the target of severe criticisms both from the scientific community and the politicians.

The scientists, wary that the station might swallow the funds allocated in NASA's budget to space science, were the most vocal. During the early budget discussions, they deluged the congressional committees with letters, urging the Congress not to cut the science budget to the advantage of the space station. They were particularly afraid of the possible consequences to the Advanced X-ray Astrophysical Facility (AXAF), the Space Infrared Telescope Facility (SIRTF) and CRAF-Cassini, a mission in which ESA had become an important partner as of 1987, when the SPC decided to include the Huygens probe in its Science Program (Figure 19).[3] Deprived of a strong scientific justification, the station was not able to create a consensus among the various scientific circles, which preferred to see the taxpayers' money being spent on smaller and less glamorous unmanned missions. Part of the Congress also wanted to safeguard the other elements of NASA's program including science, as well as other parts of the nation's budget, such as veterans' medical benefits, environmental programs, and housing for the poor and the homeless.

The first half of 1991 was particularly difficult. The cost of the station had risen continuously from an original estimate of $8 billion, when the program was initiated, to $37 billion, even after severe reductions in scope. By the middle of May, the House Appropriations Committee decided to stop the funding of Freedom. Arguing that the ''Federal government's budgeting had hit a dead end,'' they could thus redirect $1.2 billion to issues of more immediate social benefit.

At the same time, ESA was in the painful process of preparing the ministerial Council to be held in November in Munich and was striving to convince the ministers of the value of the Columbus program. There was no real advantage

Figure 19. The NASA Cassini mission comprises an orbiter of Saturn to study the planet, its rings, and its satellites; and the Huygens probe, developed by ESA, to study Titan, its atmosphere, and its surface. (ESA/NASA)

there to attacking the station. An angered ESA Director General, Jean-Marie Luton, wrote to the chairman of the U.S. National Space Council, Vice President Quayle, protesting against what he, on behalf of ESA and of the Member States which signed the IGA, considered a "manner (that) does great damage to the credibility in U.S. international cooperative commitments." He asked the Vice President "to make every effort to ensure that a level of funding is obtained commensurate with the United States' undertakings under the IGA and Memoranda of Understanding . . ." Richard Truly, the NASA administrator, reassured him that NASA was "fully and unequivocally committed to the development of the space station," and that he would "do all in [his] power to convince the U.S. Congress of the need to fund the space station Freedom program in accordance with the restructured program plans and schedule developed with our partners." He concluded by saying: "We look forward to pursuing our partnership both for its many mutual benefits and as an important step in the direction of further human exploration of space." Ian Pryke, head of ESA's Washington office, summarized the ESA view on the issue: "The U.S. must understand that the implementation of the Freedom program will be considered as a 'benchmark' against which the advisability of future large scale technical cooperation with the United States will be judged and against which U.S. claims to space leadership, on new programs, will be assessed."

On 3 June 1991, almost exactly ten years after the memorable trip to America of Vittorio Manno at the apex of the ISPM crisis, the Director General boarded the Concorde at Charles de Gaulle airport. Later that day he, together with his Canadian and Japanese partners, testified in front of a bewildered U.S. Senate. The Vice President reassured him that President Bush and he were fully engaged in the process of saving the station. He even threatened that the President might veto the bill containing NASA appropriations if the space station were canceled. The House reacted a few days later by adopting an amendment which enabled the program to continue, at the expense, however, of several other activities in NASA, in particular in space science. The final decision was left to the joint House and Senate Conference which met on 26 September.

At that meeting, NASA's budget was approved at a level of $14.329 billion, $1.4 billion short of what NASA had requested to keep its programs running. The station was fully funded at the requested level of $2.029 billion, but at the expense of other projects. The CRAF-Cassini NASA-ESA-Germany project, with a budget allocation falling short by $117 million, had to be delayed by at least one year and more likely two. This decision was—once more—made without any attempt to discuss it with ESA. Bonnet reacted with vigor and complained to Lennard Fisk, the associate administrator for space science and applications, that there had been no consultation with the international partner, whose presence in the mission had been an essential element at the time of deciding on the project in the United States some three years earlier, a fact apparently forgotten by NASA. The Director General, in a letter addressed to the administrator, protested against that unilateral decision, whose financial and scheduling implications were unacceptable to ESA. Exactly ten years after the ISPM crisis, history seemed to repeat itself. In his reply the NASA administrator cynically advised ESA to plan for a two-year delay of the full Cassini project. ESA had no other choice than to bow and to accept the decision, supplemented this time, however, by the apologies of Lennard Fisk.

And the protest was indeed not useless. At the beginning of 1992, anticipating an instruction by the new administrator, Daniel Goldin, to look at an overall reduction of the NASA program, Fisk requested the engineers at the Jet Propulsion Laboratory to reduce the scope of Cassini in order to save some $300 million out of a total budget of $1.7 billion. In the process, CRAF disappeared forever, but Huygens remained untouched. Furthermore, ESA was fully involved in the reduction exercise, which, this time, was conducted in a true spirit of cooperation and of partnership, a spirit which, since then, seems to govern the development of this unique planetary mission.

And a new President took power in January 1993. The transition between two administrations is always a delicate moment for NASA. That was indeed the case during the Carter-Reagan transition, when NASA canceled ISPM. Hence, the question everybody had in Europe, in Canada, and in Japan was what the new President would do with the space station (and with space science). Rumors were spreading that his advisors and the Congress were again requesting that the program be stopped. Indeed, they were all very critical of the whole project, which some did not hesitate to qualify as extravagant, useless, and ruining the budget, when all efforts were needed to reduce the nation's deficit. The President in his campaign, though, had been a keen supporter of the station, which secured 75,000 jobs in some thirty-seven states. This argument did not sensitize the OMB, under Leon Panetta, who recommended that the project be stopped, thereby saving $40 billion. In an atmosphere of suspense, which one day saw the station canceled and another day saw its budget reduced by 40 percent, the President eventually decided to grant the full budget requested by NASA for 1994, amounting to $2.225 billion. The acceptance, however, was accompanied by a request to Daniel Goldin to find reductions in the subsequent years and to maintain the yearly budget of the station within a limit of no more than $1.8 billion until 1998. That was clearly impossible to implement without yet another major redesign of the station.

Early in March Goldin informed Luton of the President's desire to see the station redesigned ''as part of a program that is more efficient and capable of producing greater return on investments.'' For quite some time, ESA was in the dark as to the extent of the ''redesign.'' This new disturbance was certainly not helping the Director General in his attempts to get the money from the various Member States to proceed with Columbus. European industry was ready to start developing its part of the hardware, but the present American uncertainty would delay its work for at least three more months. The only thing the Director General could do was to request that any restructuring of the project in the United States should involve the international partners. On 17 March 1993 he wrote to Goldin, bringing to his attention that both of them had ''to do [their] utmost to operate in accordance with the terms and conditions laid down in the Intergovernmental Agreement and the Memorandum of Understanding . . . that any redesign process [should be] consistent with the term and the spirit of the above mentioned agreements.'' Of concern to ESA was the possible consequence that any redesign of the station might imply for the Attached Pressurized Module. Japan and Canada addressed similar letters to the NASA administrator. The Director General insisted that it was ''absolutely essential that the International Partners be given an appreciation for the magnitude of the relevant constraints as this will signal the potential impact on the cooperation.'' And he quoted Article 15 of the IGA stating that ''each partner undertakes to make its best efforts to obtain approval for funds to meet its obligations, consistent with its respective funding procedures.'' Similar letters, even stronger, were sent to the NASA administrator by the Canadians and the Japanese, expressing their strong wish that they also be involved in the redesign process and that they be part of the independent senior-level panel (then called the blue-ribbon panel), formed by the White House to assess the redesigned options and formulate the final recommendations by the beginning of the following June.

The suspicion was growing that the station ultimately might be canceled. The science advisor to the President, John H. Gibbons, asked NASA to prepare three budget options covering the period 1994–1998, one costing $5 billion, one $7 billion, and one $9 billion (the current station program was then estimated to cost $15.1 billion over the same period). Any cost option above $7 billion would require NASA to make offsetting cuts from other parts of its programs. Some on the panel admitted that they had been given an impossible task. Reassuring words were coming from the White House that the station had to be redesigned to protect the health of American industry and to maintain the international commitments! And, indeed, the redesign team started working in good international spirit, and, in all fairness to those who at NASA were involved in the program, cooperation at the working level was going fairly well. ESA relaxed a bit—in fact, its "eagerness" to participate fully in the redesign exercise was considered by its Canadian and Japanese partners as potentially placing the integrity of the station in danger; they would have preferred a somewhat less obedient attitude.

This new episode was bringing fresh ammunition to those who, in France and Germany in particular, considered the station no longer worth the trouble and the money. The future of the program was put in question, to the great dismay of European industry, which foresaw new lay-offs of hundreds of engineers.

The story must be abandoned here. It is not finished, and no one at this stage would be taken seriously who would predict any outcome of this extraordinary saga. A majority of space officials did consider the station as already a dead project.

Following the painful ISPM crisis, the space station story would tend to show, at least after the 1991 difficulties, that NASA had understood the message, adopting a much more internationally oriented attitude. The last episodes related here suggest that the problem was no longer in the hands of NASA but had gone a step higher, to the political level of the White House, the Congress, the OSTP, and the OMB. There, the international cooperation aspects of the issue became submerged in arguments of a political nature, with internal domestic problems entering into competition with geopolitical considerations. The Russians themselves suddenly entered the scene. At some stage the blue-ribbon panel was instructed to examine the possibility of changing the station's orbit to inclinations of 51.6 degrees—the inclination of the orbit of the Russian Mir station—which would permit either a rendezvous with Mir or access by Russian launchers. The end of the cold war was marked by a curious goniometric effect.

The reader may wonder why, as scientists, we are so concerned about the fate of the station. Why are we displeased that this big swallower of space money might disappear from the programming of the major agencies? The reason is simple. We are not arguing here for or against the station on the basis of scientific arguments, but we are concerned that this program—the biggest cooperative international space venture so far—may end in a fiasco which may endanger future international ventures. We know very well that, in today's context, without international cooperation space science missions will be much more difficult to conduct, since most of them cannot be envisaged in isolation, even less in competition.

What other lessons should we draw from ISPM and the space station? Both show how difficult it is to conduct in a cooperative framework a space project whose funding requires yearly authorizations without a long-term commitment. On ISPM, ESA—perhaps naively—had expected the MOU to be a binding commitment for both parties. That was unfortunately not the case. On the space station, ESA went a step further. The Member States thought that an IGA would solve the problems Ulysses had faced before. That, again, was not the case. In fact, when it was drafted, the IGA allowed a partner to leave the project on one year's notice, but that provision was then intended to make sure that Canada, ESA, and Japan could not back out easily, not the United States!

ISPM and the space station have shown that the lack of consensus on a project makes it very vulnerable to attacks and eventually even to cancellation. So, solid widespread consensus is, particularly in the case of very expensive projects, a necessary, but not sufficient, condition. Furthermore, better and firmer cost estimates should be available at the time of project approval, or "new starts" in NASA's jargon. The cost of the space station escalated to some $40 billion from an initial estimate of $8 billion in 1984, a fivefold increase. The same trend, albeit not in absolute figures, has affected the now defunct Superconducting Super Collider (SSC) program. This procedure seems to be deeply rooted in the American system and in the budget approval process which concentrates all the limelight on the annual budget, while shifting to the shadows of the future the run-out costs. It engenders a certain uncommitted attitude, since everything can be stopped in any one year. This tendency to understate costs is the less understandable since it is well known that every four years there is a presidential election and a strong effort is then made to curb the federal deficit, increasing the danger of cuts or cancellation of cost-spiraling projects.

It may be argued that bluntly stating the best estimates of costs might jeopardize the likelihood of a project being approved as a "new start." Nevertheless, this is the procedure of ESA and its Member States. Here, projects are approved with their CaC and with provisos for participating states to withdraw from the project in case of cost increases above a certain threshold. These are two quite opposite approaches, the European system aiming at project security and strict cost control (see next chapter), the American system at project "new start" and, subsequently, defense of the yearly budget proposals.

Ideally, in the future, new mechanisms should be adopted requiring multi-year commitments. This may have to wait until the space nations have resolved their internal political problems and have in particular absorbed their budget deficits, when they will be ready to be visionary again in space. It may have to await the settlement of a new balance in international relations. For European scientists, the best partners are those for whom international cooperation represents a moral commitment, a way of behaving, and an irreplaceable asset.

6

THE NEW CONTEXT

Since the pioneering times of ESRO, ELDO, and ESA, the world scene has changed dramatically, and so has the context of space activities. Europe is now one of the most important political and economical entities in the world. European space scientists have gradually matured and little by little have occupied the front row in several areas. Ariane is the world's leading expandable system to carry payloads in geostationary orbit, having succeeded in attracting more than 50 percent of the international launch market. The success of ESA, a unique international organization, is also the success of Europe, and the future of space exploration and space exploitation is not conceivable without Europe.

Space activities have proven that they can contribute to raising the technological and industrial level of the Member States. To a certain extent, the future of European technology relies on a strong space program. Space has been, and still is, a strong element of unification in Europe, particularly through the close contacts established among the scientists, the industrialists, and the politicians. With very few exceptions, they all cite space as one of the major successes of Europe.

However, in the new international context which is now shaping up, Europe, and ESA, are facing difficult challenges, with the fall of the USSR and the increasing competition from Russia, from Japan, and also from China. ESA is also forced to analyze its future role in view of the eagerness of the European Commission to participate in the definition of the European space policy, and, more specifically, ESA must analyze the likely consequences to its industrial policy of the application of the Single European Act.

▪ THE FALL OF THE ''SOVIET BLOC''

Christmas 1991. The red flag slowly descends the mast on the Kremlin tower, and the blue-red-white flag of Russia is hoisted in its place. An event of monumental importance for those who have lived through this century, although it may be only a wink of history in a few hundred years. An event which is the conclusion of a major confrontation between different schools in the philosophical, religious, economic, sociological, and technological spheres, but one which itself will have profound implications for the future. A ''Community of Independent States'' was painfully emerging from the ashes of the Soviet Union, but it would take time before an ordered system might appear, ensured of some stability.

The space program of the USSR has, beyond any doubt, been one of the greatest technological achievements of that country, if not the greatest. It has imprinted its mark on the space era and has had a powerful influence on its orientations. Those of us born in the 1940s and before remember our stupefaction when we heard the bip bip of Sputnik-1, the first manmade satellite in orbit, our sorrow and compassion for Laika, the first living creature in space, and our awe and incredulity when Yuri Gagarin, a man like us, flew for the first time in space. Who, at that time, was not a fan of Yuri? These achievements had tremendous implications for the technological effort of the other superpower, which accepted and admirably confronted the challenge . . . and who, in 1969, was not a fan of Neil Armstrong? Then, of course, after that climax, the public quickly redirected its sights to more terrestrial things and only occasionally lifted its attention again to space such as when the Challenger exploded or when Buran, the Soviet shuttle, made its maiden flight.

The USSR space program was an indisputable success achieved in conditions difficult to imagine, when photocopying machines were still suspected of being enemies of the working class. Later, with the USSR sailing into increasingly rough economic seas, but with increased *glasnost,* the space program came under criticism. This led Roald Sagdeev, the long-standing director of the USSR's Space Research Institute (IKI) and, until he left for the United States in 1988, the personification of the Soviet space program, to say in 1990: ''In the past, we would have said about the space program that the present is obscure but the future is brilliant. Nowadays we would say that the present is brilliant but it is the future which is obscure.''

Now everything is indeed obscure and a matter for political speculation. Paradoxically, the fairly successful science program of the USSR is now in jeopardy. Owing to severe budget cuts, the Mars-94 mission, which comprised the launch of two identical satellites to Mars in 1994 and the landing on the surface of balloons developed by France, was first reduced to a single launch in 1994, with no balloons, possibly followed by a second launch in 1996 carrying the balloons and a rover; later the 1994 mission was postponed to 1996. The future of that program relies on whatever charity the involved Western European space agencies—including ESA—might agree to grant. Budget cuts did put an end to the Cluster/Regatta mission, a cooperation between IKI and ESA.[1] The European high-energy astrophysicists, who enthusiastically jumped at the opportunity offered to them in 1987 by the very ambitious Spectrum X-gamma Soviet observatory, are now quite concerned that their efforts and investments may have been in vain, and that they may have to wait many more years before they can analyze the first X-ray and gamma ray photons collected by that mission.

The Soviet space program has been criticized for its inability to cope with the ecological disasters which plagued the country and to properly survey the USSR territory. The government has reacted by setting up commissions on the environment, and a move was made in the late 1980s to shift from a purely military Earth-observing program to a more environment-oriented one. The Russians now show an unusual spirit of openness and a keen and genuine interest in cooperating in the sensitive area of remote sensing.

Surprisingly, the manned program, which in Europe and in the United States is under severe attack, is maintained in Russia at a high level of activity, with the Mir station permanently occupied and opened to utilization by an increasing number of cosmonauts from many different countries. However, the mighty Energia launcher, after its first two successful flights, is no longer used and Buran, the space shuttle, at times highly criticized by Soviet scientists, including Roald Sagdeev, still has no clearly identified utilization and runs the risk of being abandoned in the rusty superstructures of the Baïkonour Cosmodrome. The future seems foggy to say the least, in spite of the creation of a Russian Space Agency in 1992, which has led to some stability and increased security in the space program and has considerably simplified the interface with the space agencies of the world. Under these circumstances, on the one hand, it may seem delicate to embark at this moment on a program of a certain scope with Russia. On the other hand, it is of interest to take advantage of the enormous space capabilities inherited from the strong Soviet space program. In this new context nobody can predict for sure how space activities might develop in every part of the world, except that the main engine—the cold war—is no longer pushing the space effort. In space science there seems to be no other issue than to cooperate. In particular, the role that the Russians might be able, or will decide, to play will have a strong influence on the future of space activities around the world. Several do not look with favor on an era dominated by an unchallenged U.S. leadership.

What can ESA do in this context? The ministers who met in Munich in 1991 and in Granada one year later have indicated the way. The best course is obviously to keep open all options of cooperation in the area of space physics, Earth observation, life sciences, and manned space flights, using the unique Mir station. ESA should even be prepared to embark on more ambitious joint missions with the heirs of the Soviet program. If it does not, others will do it instead. It is therefore obvious that a coherent and visionary approach to cooperation with Russia ought to be adopted.

▪ CENTRAL EUROPE

As a consequence of the elimination of the USSR, all former socialist countries are impatient to join ESA. None of them, however, is able to catch up immediately and without some preparation to the already existing Member States. All of them do possess very highly qualified scientists and engineers, but they cannot yet compete with success on the technological and industrial front. An accelerated and regular growth of their economies, with special efforts to foster a competitive industrial capability, might shorten the waiting period. Therefore, some special agreements must be found. For instance, they can be users of the services offered by ESA or negotiate, as other countries have done, special contracts of association.

These countries are culturally and historically part of Europe, representing an overall population of some 120 million. They cannot be ignored, even though their full integration into ESA appears today as a fairly remote possibility. Their respective scientific communities have a long tradition of excellence, and it is just a matter of time before they may be in a position to take full advantage of their participation in the ESA system. It is in the interest of Europe to integrate them fully into its political and economic system.

The policy for ESA will therefore be to gradually open its programs, while encouraging these countries to develop a substantial national educational effort and to establish contracts of association in space and in Earth science, meteorology, telecommunications, and possibly space technology. Connecting various institutes or laboratories and archival centers through information and data networks offers some interesting possibilities. It may also be efficient to allocate each of them every year a number of fellowships, in order to create a nucleus of space scientists and engineers educated in the ESA Member States, provided they return to their mother country at the end of their fellowship. Certainly, these measures are not sufficient and can work only if completed by national policies with the aim of increasing the rate of growth, leading to a progressive integration into the ESA community.

It would indeed be imprudent—and irresponsible—to make these countries believe that the process of full integration into ESA can be immediate. ESA's experience has shown that it is wise, for the smaller—or the less rich—countries, to proceed gradually toward this goal, and to join ESA fully once their technical and scientific achievement makes them competitive with respect to their more senior partners. The European political context may force an acceleration of this process. Europe, and ESA, should tackle this challenge. In the present situation, specific and special association agreements are more realistic than full membership. Also, regional initiatives may help in the meantime, like the Central European Initiative sponsored by Italy and Austria concerning the development and launch of a small regional satellite (CESAR) together with Hungary, Poland, and the Czech and Slovak Republics. These initiatives are more likely to bring positive results to both sides—and to do so faster.

■ ESA IN A BORDERLESS EUROPE

Following the ratification of the Maastricht Treaty by the various European Union Member States, the entering into force of the Single European Act may yield the EU a stronger role in space matters. Some ESA countries which up to now have not been members of the Community, like Austria, Sweden, Norway, and maybe soon Finland, will join the EU, and some EU countries, like Portugal and Greece, are expressing some interest in entering ESA or in being associated with it. Thus the difference between the EU and ESA membership will be reduced in the future. Therefore, it is legitimate to ask whether ESA could maintain its cohesion and stick to its role, as spelled out in its Convention, of defining and implementing the space policy of Europe.

Indeed, the EU is showing a growing interest in the matter. In 1988 it published a document describing its proposed involvement in space activities. Three years later an expert panel working for the Commission and chaired by Roy Gibson, a former ESA Director General, claimed that "Space is too important to be left to space agencies." This spectacular statement was made one month before the ministerial meeting of November 1991 in Munich. Gibson's panel considered that the EU should have a role to play in defining—and implementing—the future space policy of Europe, a responsibility solely assigned to ESA up to now. There are, however, basic differences between the EU and ESA.

The former claims a real transfer of sovereignty, from the States to the Commission. ESA does not: it maintains in existence both a common European program and national activities. The European politicians have created an ESA, not a European NASA. The EU aims at establishing economic and political unity. ESA is essentially a technological development agency. Indeed, the Commission has competence in research and technology matters, of which space activities are an important element. In the panel's report, Gibson urged the Commission to prepare Europe's space industry for an increasing world-wide competition and recommended, wisely, that the Commission and ESA better coordinate their research and technology activities. The Council of the EU research ministers, meeting in April 1993, agreed to make a stronger effort to incorporate space research into a much wider range of the Commission's own research and development programs.

There are many issues associated with the future relationship between ESA and the EU. The Commission is certainly interested in the relevance of the Earth observation program to a better implementation of its own policies in agriculture—the Commission is Europe's largest customer for ESA's remote sensing images and involved the equivalent of $12 million in European remote sensing industries in 1993—as well as its policies concerning the environment and development aid. Generally speaking, it sees its role as including the space component in the implementation of its own policies. It is keen to widen as much as possible an open industrial competition, and therefore may not view very favorably the continuation of a strict application of the *juste retour* principle on which ESA's industrial policy is based and which has been one key element of its success. Some coordination is already under way in the areas of technology and environmental policies. But these will require more and more attention in the future. Both Director Generals, Reimar Lüst and Jean-Marie Luton, have met with Jacques Delors, the President of the Commission in Brussels. ESA's Council, for its own part, has established a special working group to analyze the effect of the Single European Act and its consequences on ESA policy. A great part of their analysis rested on the interpretation of the legal rights of both ESA and the Commission to apply each other's rules, as defined in the Convention and in the Rome Treaty.

The implementation of the *juste retour* principle at ESA is certainly a priv-ileged target for the Commissioners. The Single European Act aims at the creation of an ''area without internal frontiers in which the free movement of goods, persons, services and capital is ensured.'' It aims at ending the national character of markets within its Member States, and at prohibiting the states from aiding national undertakings. ESA's industrial policy favors the awarding of contracts nationally and preferentially within each Member State contributing to a given program. But the EU advisors considered this policy as hampering competition and decreasing the overall competitiveness of European aerospace companies. At a minimum, they recommended that ESA go back to its Convention, applying the *juste retour* rule globally and not program by program.

One could argue, however, that ESA contracts are not national procurements, since they are placed by an international organization acting within the scope of its Convention. One can also argue that ESA procurements are research and development contracts and, as such, do not distort the trade rules which govern the application of the Single European Act, essentially concerned with the commercialization of developed products. The problem is therefore where to draw the line between research and development activities on one side and commercial activities on the other. Certainly, the Rome Treaty, which governs the EU policy, foresees cooperative agreements with other international organizations in the area of research and development. Even though ESA, as an international organization, is not bound by EU laws, the problem for Member States which are part of both ESA and the EU is to know whether they are themselves bound by the EU laws, whether they are allowed to participate in ESA and to implement its policy. The Convention offers the possibility of national procurement rules, while the EU open competition approach may forbid the Member States to continue applying their somewhat protectionist policy. Some Member States have threatened to withdraw from ESA's activities if they are forced to apply the EU policy to the placing of space contracts. Indeed, many of the smaller companies, which are protected by the present ESA policy, might be wiped out in a worldwide competitive market if they could not get a fair return. If this were the case, the implementation of the Single European Act would have effects contrary to its own objectives.

Worthy of notice is the reaction of industry itself to the prospect of relaxing ESA's industrial policy. Thirty-six European companies, representing 92 percent of the European space industrial turnover, participated in an enquiry conducted by Eurospace, an association of most European aerospace companies. From this enquiry, a clear distinction could be drawn between the countries hosting the prime contractors—the larger Member States—and those which participate only at the subcontractor level. While the former had no difficulty abandoning the fair return principle, the latter, in their great majority, were strongly in favor of sticking to it. These same companies were at the same time opposed to free competition, while those of the larger ESA countries had no objection to it, but they could accept a compromise in order to protect the European market against the penetration of non-European industries. Hence, it should not be too surprising that the large companies are in favor of an active role of the European Commission in the sector of telecommunications and remote sensing, while the small companies are of the opposite opinion, all agreeing, however, and contrary to Gibson's recommendations, that in the fields of launchers, space stations, and science the Commission should not intervene at all. On the other hand, the regrouping of firms into European multinational industrial entities may relax the pressure from the smallest Member States to adhere too strictly to the fair return principle. Industry contends that neither the EU nor ESA is helping them enough in the area of telecommunications, where they fear a fiercer competition when the European market will be deregulated and flooded with the cheaper U.S. and Japanese equipment.

As we can see, the implementation of the Single European Act has already induced a large amount of reflection. If the EU and ESA memberships are exactly the same, one may wonder whether ESA should continue to be governed by the same rules as before. The supporters of increased competition might favor a new approach. However, the positive effect which the space program has had, up to now, in improving European integration, may be adversely affected by the increased dichotomy which may develop between the larger and the smaller countries. Hence, it is essential to ensure that the efforts of ESA and those of the Commission ultimately promote overall European interests.

In this context, a stronger coordination between the two institutions is desirable. If a common ground of understanding could be found, it would shift the space commitment of the Member States from an essentially industry-oriented motivation to a more political one, an evolution which might have the positive effect of ensuring a larger role for Europe in the future exploitation and exploration of space. Two areas in particular deserve special attention: the launchers and the environment.

In the launcher area, there is some hope that wisdom might eventually prevail, as evidenced during the discussions held in 1992 between the Americans and the Russians on the commercialization of the Proton launcher which, due to the very low price offered by the Russians, placed Europe's own launcher,

Ariane, under severe threat. ESA and the EEC together, in a well-coordinated approach, insisted that Europe should be present at the negotiating table, and both participated. A clear separation of their respective responsibilities and a close coordination of their respective policies, based on a good understanding of the overall European interest and of its advanced technological and scientific achievements, should avoid potential conflicts of interest and minimize overlap.

In the area of Earth and environmental sciences, the lack of a reference program in Europe has opened the way to a number of uncorrelated initiatives. The creation of a European Environmental Agency, whose role is not fully clear, under the auspices of the Commission, is one of them. ESA would be a stronger partner had it established a program similar to Horizon 2000, supported by the scientific community. Because such a program did not exist, national projects were initiated in France, in Germany, and in Italy, not necessarily leading to the best utilization of the overall European resources. Fortunately, some coordination was secured through bilateral discussions between the participating countries.

Also, discussions have been initiated to define a European space program for verification of disarmament. National initiatives are being taken in this very sensitive domain, and ESA should consider whether it is bound forever to its peaceful-only role. The interest in the observation of our planet should not lead, however, to a situation in which the scientific character of most of the problems linked to the global change of the climate and of the environment is made subsidiary to the necessities of more applied or commercial and strategic objectives.

▪ MUNICH, 20 NOVEMBER 1991

Those who believe in symbols could have guessed that the third ministerial meeting, after Rome in 1985 and The Hague in 1987, would be a difficult one. In January 1985 the Sun was shining on Rome through the cypress trees of the Villa Madama on the hills which overlook the Tevere river. In contrast, in November 1987 the Sun was hidden by a chilly fog which hung above the canals in The Hague. But when the ESA delegates landed in Munich on 18 November 1991, they found the Bavarian city covered with snow and the air very cold.

Indeed, the delegates in Munich had a difficult task in front of them. In Rome there had been little to decide, apart from the budget increase of the

Science Program; the ministers had simply to lay down their plans for the setting of the In Orbit Infrastructure, leaving any commitment on the related budget to a later appointment. The Hague meeting was a little less easy. A decision was made there to continue the studies on Hermes, the European space plane to be launched by Ariane-5, and on Columbus, the European participation in the space station, while the Ariane-5 program was authorized to proceed fully into its development phase. In Munich, however, the ministers had to agree on the full bill.

For more than a year ESA had been preparing that meeting. The difficulty was to reduce the overall costs of the Infrastructure (Ariane-5, Columbus, and Hermes) by some 15 to 20 percent, in response to a request by the Germans, suddenly confronted with the economic strains of their reunification. In spite of its efforts, the Executive was unable to achieve better than a modest 11 percent reduction over three years. Reducing further would have led to unacceptable delays, themselves inducing additional costs, and to a disbanding of the complex industrial distribution of tasks which had been agreed upon after long negotiations with all the Member States, among which the smaller were not the least interested in an early start of the program.

Since The Hague, major changes had occurred in the world which directly influenced the balance of Western Europe: German reunification, the dismantling of the Soviet bloc and breakdown of the Soviet Union. The immediate effect had been a slowing down of the economies of the Member States, resulting in a reduction of the resources allocated to research, and leading, in the space program, to a slower pace of work and a stretching of the industrial tasks necessary to implement the various elements of the long-term plan agreed upon in The Hague—the overall spending had been some 24 percent lower than envisaged originally. In addition, the growing importance of environmental issues, the greenhouse effect and the hole in the ozone, had focused the attention of the politicians and dictated new priorities. This was the context in which the Council, its working group, and the Executive had for more than twelve months prepared the Munich meeting.

Once more, France and Germany held in their hands the fate of the delicately balanced compromise agreed upon by all Member States only a few days before the meeting. The two countries had been arguing about Hermes and Columbus, their respective pet projects. The bill, however, even after the reduction efforts of the Executive, was still considered too high for the host country. It was hoped that a compromise could be reached on the occasion of the French-German summit to be held in Bonn on 14–15 November, just four days before the opening of the ESA ministerial meeting. There, the German Chancellor and the French President, indeed, had been keen to reach an agreement on the future of Europe, paving the way to the establishment of the Maastricht Treaty.

The Council, chaired by Claudio Aranzadi, Spanish Minister of Industry, Commerce, and Tourism, met in the Kaiser room of the Residenz, the palace of the kings of Bavaria. The ceilings and the frescos which richly ornamented the room symbolized, among other things, the figure of generosity. What better symbol could have presided over the meeting? Unfortunately the symbol had little effect, and the consensus was long and painful to formulate.

After two difficult days of discussion and painful negotiations, in large part held in private rooms and corridors, on 20 November the delegates eventually managed to endorse unanimously—although for some of them reluctantly—two resolutions: one on the European long-term space plan, covering the period 1992–2005, and one on the observation of the Earth and its environment. They requested, however, the Executive to achieve a further saving of 5 percent on the 1992 budget, with improved cost estimates for Hermes and Columbus. This came as a surprise to most delegations, which had counted on the French in particular to exert strong pressure against the cost-reduction efforts of the Germans. But the compromise agreed upon in Bonn a few days earlier by the French President and the German Chancellor did not leave much freedom to the French Minister, Paul Quillès, to negotiate an immediate start of the manned elements of the long-term plan, in particular Hermes, which the Germans still considered much too ambitious.

Furthermore, the ministers, "recognizing the need for a careful, ongoing analysis of the changing geopolitical context, in order to assess its impact on European space activities," reaffirmed the need to intensify international cooperation, both among the Member States and with other European and non-European partners, in particular Russia, in order to carry out the long-term plan with maximum effectiveness and at the best cost. They fixed the date of the following meeting for the end of 1992, to be held in Spain.

The consensus finally reached at the Munich meeting was hard for some of the Member States to accept. It was a cause of concern, especially for their industry, which had expected that Munich would give the go-ahead for the development of the major elements of the long-term plan.

Were there really cost savings to be expected in opening the programs to new, but not necessarily rich, partners, such as the Soviet Union, which at that very moment was rapidly falling apart, or the countries of its former bloc? Reimar Lüst, the artisan of The Hague compromise, could not hide his disappointment. He doubted that cooperating with the Soviets, in the present state of disarray of their country, would lead to substantial savings, if any. Several other delegates shared this view, in particular those in the smaller Member

States, who foresaw with displeasure large portions of industrial work leaking out to Russia.

However, there was some reason for hope. For example, the ministers expressed satisfaction with the success of the Agency, in particular in science and Earth observations, which were symbolized by the spectacular views of our planet obtained by ERS-1, a mock-up of which was sitting in a place of honor right beneath the windows of the Kaiser room, in direct sight of all the ministers and delegates. But something indeed had changed in Munich in November 1991. Back in Paris, the Executive looked at its torn-apart long-term plan and returned philosophically to the drawing board.

■ GRANADA, 9 NOVEMBER 1992

The Sun was shining again over the hills from where the Alhambra dominates Granada, the city selected by the Spanish authorities to hold the next minis-terial council. With only a year between the two meetings, the ministers had very little time to change their minds, and ESA little time to prepare the meeting. The Director General had formed a "strategy group" under the responsibility of Jean-Jacques Dordain, a French engineer from the Office National d'Etudes et de Recherches Aéronautiques. The group worked full time, full speed, preparing scenario after scenario. The dilemma was whether to propose a plan fitting the less ambitious budget that the Member States could afford, therefore with less ambitious objectives, or a plan that would stick to the Munich objectives but that could not fit the budget.

The delegates were hesitant to tailor the objectives too tightly to the budget. They thought, probably rightly, that this would not be very attractive for their ministers. A long-term and, if at all possible, ambitious vision seemed pref-erable and would more easily convince the various governments that it was worth committing funds to the space venture. But how to maintain the objec-tives within the financial realities? Dordain and his group did not spare their efforts or their imagination, but they had a hard time finding their way in between these narrow borders. There could be no miracle. A realistic estimate of the financial resources of the Member States led to the inescapable conclu-sion that one of the elements of the Munich plan had to be abandoned or indefinitely delayed. What had been possible to accept and finance a few years earlier was now beyond the resources of the Germans and the French. Stretching the program would not solve the problem as it would induce extra costs. The new budget reality imposed its new and restrictive law. Which would be the victim? Hermes or Columbus? France or Germany? And what about international cooperation? Was there any realistic hope that the Russians, in particular, would be able to contribute to the solution of this dilemma? The questions were wrestled with, meeting after meeting, around the table of the Council at the ESA Seat in Paris.

In the meantime, the financial situation of France and Germany had not improved. Hubert Curien, who had succeeded Paul Quillès in 1992 as Minister of Research and Space, seemed much less adamant than his predecessor to pursue the ambitious goals. At the same time, there seemed to be no hope that the Germans could adapt their resources to the objectives outlined in Munich. Not only Hermes had to be drastically redesigned, but also the various parts of the Columbus program, so dear to the heart of Heinz Riesenhuber, the German Minister. Curien, confronted with the problem of reducing the overall French budget deficit in view of the upcoming parliamentary elections of March 1993, admitted that his government could not pursue Hermes, to the great dismay of Jean-Daniel Levi and Daniel Sacotte, the two top-level officials from CNES, who had been fighting hard to keep the space plane in the program, and were striving to give Europe the ambitious long-term plan which they thought it deserved. The Germans, in turn, had to admit that the Man-Tended Free Flyer, that symbol of European independence, could not be continued and had to be abandoned. The grandiose visions of The Hague, so severely criticized at the time by the U.K. Minister, Kenneth Clarke, had not resisted the new political and financial situation of Europe after the cold war.

This sudden change of attitude was frustrating for the smaller contributing countries, which had always considered France a leader, the source of inspiration of the whole European space program, and, as such, the carrier of the hopes of their own national industries. Belgium, in particular, a long-term space ally of France, was bitter and expressed disappointment with an emphatic defense of European cooperation. Spain was not pleased with the idea of subsidizing Russian industry, although this was the only realistic way to learn how to cooperate with Russia. However, little by little, realism and reason came back again to the table of the Council.

Until the last moment, hectic negotiations took place, involving the Executive, the President of the Council, the Spanish delegation, whose country was hosting the meeting, and Hubert Curien, who was chairing it. In fact, when the ministers met in the Congress Center of Granada, all had more or less agreed on the content of the proposal. The meeting therefore devoted most of its time to what seemed to be subsidiary matters, such as the problem of the retroactive adjustment of contributions, a question of particular sensitivity to the Member States which, like Italy, the United Kingdom, and Spain, had just gone through a devaluation of their currencies, and as a consequence had to pay more money to ESA in order to honor their contributions evaluated in Accounting Units. Not too surprisingly, also, the meeting spent a considerable amount of time discussing the sacrosanct issue of industrial return. The ministers took advantage of the meeting to, once more, raise the coefficient below which special measures ought to be taken, from 95 percent to 96 percent over the 1994–1996 period.

By the end of the meeting a consensus had emerged, though a painful one for several Member States which foresaw a severe reduction of the workload of their industrial teams. The Columbus Man-Tended Free Flyer had disappeared from the plan, and the studies of Hermes were ''reoriented'' for a period extending from 1993 until 1996, according to three cooperative options: one with Russia, one with the United States, and a Europe-only scenario. As they had in Munich, the ministers emphasized the importance of the Earth and environmental problems. Hence, they decided to initiate the Columbus polar platform program, starting with two polar orbiting satellites: Envisat-1 to study the environment, and Metop-1 for operational meteorology. As in Munich, space science was untouched, but the ministers requested the Director General to present a future long-term plan for European space science in 1995, when they would meet again.

Probably the most important political decision made at the meeting bore on cooperation with Russia, and on the joint study of a possible successor to the Hermes program. By this decision, in a sense, the ministers opened the way to some globalization of the space activities. This measure was supported financially with the authorization given to the Director General to allocate 110 MAU worth of contracts to Russian industry for joint studies. The goal of reducing the costs to Europe of the big infrastructure programs, as recommended in Munich, was certainly not achieved. But could it be achieved, given the situation of Russia, with nearly all its space capabilities falling apart?

Munich in 1991 and Granada in 1992 were a turning point in the history of ESA and of international space cooperation in Europe. ESA, which had aimed at being a modest third partner behind the United States and the Soviet Union, was promoted to number two after the fall of the USSR, but would it be able to honor that position?

▪ HOW TO COOPERATE IN THE FUTURE?

At the Granada meeting, the U.K. minister, to the surprise of several of his fellows around the table, recommended that the Council support the idea of a World Space Agency, which would be the proper forum for the broad international cooperation which should govern space activities in the future. Some were surprised and others smiled at the concept, which had been proposed several times in the past on many different occasions by several different politicians. The Council took up the point, however, and did mention that the IACG for space science, and the Space Agency Forum, a group of nearly thirty space agencies and related organizations in the world, formed in the framework of the International Space Year, were examples on which to frame the future. Clearly, the idea of a globalization of all space programs was latent in Granada.

The future of Europe's international cooperative ventures will depend, first, upon the existence of a strong European space program, under the aegis of ESA. Hence, on the eve of the next millennium, confronted with a series of new challenges, in a world which has evolved beyond all expectations, ESA has to reanalyze its plans and must assess how it is now possible for its constituent Member States to cooperate in the most efficient way and on what programs.

All space programs seem to grow faster than the funds allocated to implement them. Hence, the future in space will be marked, by necessity, by an increasing international cooperation. The major space programs will most probably be mutually dependent. For Europe, the United States, Russia, Japan, and their partners, there are very few other alternatives than to cooperate, but one important issue will be for them to accept this mutual dependency. As to the United States, it should become a more reliable partner, through faithfulness to long-term commitments and adherence to international agreements. It must admit that the defense of its ''leadership'' is not necessarily an objective of great appeal to its partners and that it should adhere more often to the concept of ''partnership.'' Sometimes the United States may lead and sometimes not. As Ian Pryke, the head of ESA's Washington office, says: ''The right to lead must be earned by today's actions and not be claimed by reference to yesterday's achievements.'' As to Europe, it should learn how to speak with a voice that reflects a better-coordinated approach, after having decided what objectives it intends to follow. The Europeans should understand that they have matured, that it may soon be their turn to drive the system. Only if they are determined and ambitious enough will they become more credible and more respected partners. The Russians, for their part, should reach a more stable situation and soon stop relying on the other nations' charity. However, the future evolution of their space program is still overshadowed by political, economic, and organizational uncertainties.

All, Europeans, Americans, Japanese, Russians, have something to change in their present practices if they want to continue exploring space with ambition and vision in the next century. Sticking to the present state of business might well limit the future of world space activities to a set of rather modest ventures, undertaken with poor financial efficiency, or in isolation.

In this respect, we cannot insist too much that the modes of participation in the space station are a test for any future venture of a similar scope, such as a return to the Moon, a mission to Mars, or any other science mission of the same ambition. Should the space station fail because of decisions made with no spirit of partnership and no consideration of the budget or political aspects involving all partners, it would be difficult for the partners to embark soon on any other similar venture. However, maintaining such big infrastructure programs cannot be done at the expense of other missions, because, in that case, the whole support of the users community and of the general public may be lost. The success of ESA's participation in Russia's manned program as recommended by the ministers in Granada is also a test for future ventures. However, within the limits of its relatively modest means, if ESA can hope to participate, it cannot have a driving role in the future possible exploration initiatives. It is nevertheless important that it be prepared to participate and respond to any realistic and advantageous opportunity which may arise in the future.

ESA must also adapt itself to the evolving situation of today's Europe, watch carefully how its main partners will reorient their partnership in the context of this changing world, and openly and constructively face the challenge posed by the increasing importance of countries like Japan and China. For this to succeed, however, the highest level of political and financial commitment and the determination of its Member States are essential to at least maintain and, better, improve, the leading position it has now reached in several sectors after thirty years of existence.

CONCLUSION

In the overall development of space activities, ESA appears as a unique example of its kind. One may therefore wonder whether ESA might not serve as a model for international cooperation on an even larger scale. Looking backward, ESA has been indeed quite successful. This can be judged on the number of science and application missions launched since the ESRO days (Tables 2 and 3), a number which is certainly modest when compared with what the Americans and the Soviets have launched since the beginning of the space era but which, with very few exceptions, includes no major failures and plenty of successful missions. On that basis, but also looking at the high scientific and technical level of these missions, ESA, without any doubt, is one of the top three space organizations in the world. A closer look at its achievements shows that it is even leading in several sectors. Ariane, the European launcher, has attracted more than 50 percent of the commercial market. The breathtaking performance of Giotto in its momentous encounter with Halley's Comet in 1986, and its successful flyby of Grigg-Skjellerup in 1992, at a distance from the nucleus of only 200 km, will remain in the history of space exploration. ESA has successfully developed an operational meteorological program which is a key element in the success of weather forecasting in Europe, and even in the United States (in 1993, Meteosat-3 has been moved 75° westward to cover the United States, replacing a non-operating NOAA satellite). ERS-1, ESA's first Earth-observing satellite, with its powerful synthetic aperture radar, is observing the Earth twenty-four hours a day and through the clouds: it is also an indisputable success. Even Spacelab, although it is no longer in the hands of ESA, is the only comprehensive and versatile space facility capable of performing, during the same flight, experiments in microgravity, astronomy, and geophysics.

This success cannot be dissociated from the strong political determination of the Member States to join their efforts and Europeanize their programs through ESA. Without their political support, ESA might not still be in existence, and we can only hope that this support will be maintained and that the Agency will be able to survive in this changing world. It also owes its success to the existence and proper implementation of an industrial policy which has benefited the Member States' technological development and has contributed to enlarging and improving their industrial base. Of course, for states to be part of the Agency and take full advantage of their participation, it was necessary that they already possess an industrial substrate on which the industrial policy could be implemented. Obviously, not all the countries of the world do yet have such a substrate, and, if ESA can serve as an example, it cannot be generalized at this time.

ESA is also a unique and successful example because, from the stage of a research and development agency, it has created its own pre-operational entities which are now supplying services to Europe in the broadest sense. Thus, Arianespace commercializes the Ariane launcher. This is the first private company of its kind, and very successful, as we have seen, in spite of increasing competition with the United States, and now with Russia and the Chinese. Eumetsat is currently operating ESA's meteorological satellite Meteosat, and is distributing meteorological data to most European national meteorology offices. Thanks to this service, the quality of weather forecasts in Europe has considerably improved since 1980. Similarly, Eutelsat is providing space telecommunications, broadcasting, and high-definition television to nearly all European countries. Finally, Europe is participating in Inmarsat, a sixty-three-member organization, which operates satellites for the improvement of maritime communications services, with an involvement of European industry amounting to approximately 67 percent of the overall effort.

ESA has also managed to maintain a good balance between mandatory and optional activities. Obviously, the Agency could not survive with only a purely mandatory scientific program as in ESRO days. This is in fact why ESA was born. But, conversely, would ESA have survived the various crises which have marked its history so far, if its activities had been of a purely optional character? An agency with a limited scope, with a specific and time-limited mandate, can certainly operate within a purely optional framework. But an agency with a more permanent mandate needs stability and has to establish rules which secure continuity in its activities. The Mandatory Program, the Science Program, is often called the backbone of ESA: it provides the Agency with the stability it requires to conduct all its activities, whether of a mandatory or an optional nature. Without the Mandatory Program it would be more difficult to associate permanently all the Member States together. Conversely, it was in the framework of the increased number of optional activities that the science budget could benefit from a substantial and unprecedented increase in 1985, thanks also to the strong support given by the scientific community to the Horizon 2000 long-term plan. However, especially after the 1991 ministerial meeting in Munich, one may wonder whether the concept of optional programs is not now reaching a limit, by placing the fate of infrastructure programs, and consequently the stability of the Agency, predominantly in the hands of the largest-contributing Member States. Is ESA, after Munich and Granada, not getting close to a situation of ''fission'' as a consequence of this unbalanced structure, like a too massive and unstable atomic nucleus? Only the future will show whether this is true.

Certainly, creating ESA and leading it to where it is now was not easy. In this book we have addressed some of the difficulties encountered: political, industrial, and others. The ESA we have described remains an example, nevertheless. A unique example. That of a collaborative and unified effort in space among a large number of sovereign countries. By broadening its international base, opening its programs to Canada and Finland and maybe, soon, to other countries, ESA can show how it is possible to attract an even larger number of states, creating the first nucleus of a globalization of space research.

As imperfect as it may be, ESA is the only organization that has proven that it is possible for many nations to work and plan together in space activities. It is offered here as an example on which to reflect. ESA managed and succeeded. The proof is there to be seen. The model exists.

NOTES

BIBLIOGRAPHY

ESRO AND ESA SPACECRAFT

THE HORIZON 2000 LONG-TERM PLAN

ILLUSTRATIONS

INDEX

NOTES

1. The Birth of ESA

1. In 1959 the scientific advisor of NATO proposed to create a sort of European NASA to coordinate the initiatives of the European nations wishing to cooperate with NASA. In his capacity as president of the executive council of the Committee for Space Research (COSPAR), created in 1958 under the umbrella of the International Council of Scientific Unions, Henk Van de Hulst, of The Netherlands, was invited by NATO to describe the plans of COSPAR for coordinating space research at the international level. His position reflected the reluctance of European scientists to pursue such an initiative. The matter was then closed and no longer discussed in NATO circles.

2. The accounting unit at the time of ESRO was based on a gold standard in order to ensure equivalency with the U.S. dollar (1AU = $1). In 1975, however, the Council voted to make it equivalent to the standard unit of account created by the Council of the European Communities; since 1980 the accounting unit has been in principle equivalent to the European Currency Unit (ECU).

3. Following a review of the management of ESA's scientific program in 1989–1990, under the request of the U.K. Delegation, a group of consultants recommended that the Director General assess the possibility of moving the Directorate of the Scientific Program back to ESTEC. Their analysis showed it to be cost-beneficial. The move was not implemented, however, because it would have isolated the top management of the Science Program from the other Directorates inside ESA and in particular from the political decisionmaking bodies or administrative departments, the great majority of which are located at the ESA Seat in Paris.

4. This proposal was strongly opposed by the first Director, Ernst Trendelenburg, and in February 1965 ESLAB was established in Holland at Hotel Helmhorst in Noordwijk. It moved into prefabricated buildings on the ESTEC site a few months later. After the fire in 1966, ESLAB moved to another hotel in Noordwijk, Hotel Zinger, and then to a new building in the nearby village of Noordwijkerhout, where it remained until April 1969, when it moved to ESTEC. In September 1968 ESLAB was officially fused with ESTEC and renamed the Space Science Department (SSD).

5. Lüst remained in this post until the formal establishment of ESRO in 1964. The post was vacant until 1965 when Bert Bolin from Sweden was appointed.

6. In the present ESA organization, the roles of these two committees are played by the Scientific Program Committee (SPC) and the Administrative and Finance Committee (AFC).

7. ELDO decided in 1966 to build its own launching base on the grounds of the Centre Spatial Guyanais operated by CNES at Kourou in French Guyana.

8. The respective financial shares of the three main contributors were 38.79 percent for the United Kingdom, 23.93 percent for France, and 22.01 percent for Germany.

9. The ten were Belgium, Denmark, France, Germany, Italy, the Netherlands, Spain, Sweden, Switzerland, and the United Kingdom. Ireland and Norway were accepted as associate members; they became full members in January 1987, as did Austria. Also in January 1987, Finland became an associate member, aiming at full membership in 1995. In 1981 Canada signed an agreement of cooperation. At present, therefore, ESA is a thirteen-member organization with two other states participating in its activities.

2. *Governing Principles*

1. Hipparcos is the first space astrometry mission ever undertaken. It aims at measuring the positions and distances of stars to an accuracy of two thousands of an arc-second. Hipparcos was launched in 1989 with an Ariane 4. Even though the failure of its apogee boost motor did not place Hipparcos on the right orbit, the spacecraft has accomplished its nominal mission fully and successfully. The mission ended in August 1993 when the on-board computer stopped operating under heavy particle bombardment in the Van Allen belts.

2. In space, the main world market competitor is U.S. industry. However, Japan, China, and Russia are now challenging European and U.S. industry, particularly in the launcher area. Russia expressed, in 1992, the intention to commercialize the Proton launcher at a very cheap price.

3. There is no such principle as a "science return" which would guarantee every Member State a participation of its scientific community in proportion to its contributions to the Mandatory Program. The participation of scientists is secured by the quality of their proposals selected through an open competition and not by fixed quotas.

4. The implementation of a space mission is usually split into phases which correspond to contractual milestones, or which involve contractual commitments. For scientific missions in ESA, five phases are generally considered: the assessment phase, in which broad scientific objectives are converted into a set of specific performance requirements; Phase A, which involves industry and during which performance requirements are converted into a set of system design requirements; Phase B, in which the system design requirements are converted into a specific design; Phase C/D, in which all parts of the system are converted into hardware; and Phase E, which starts when the spacecraft is in orbit and corresponds to the operation and utilization of the system.

5. The prime contractor is responsible for managing all activities in industry that are associated with the development of a project. It coordinates all the subcontractors in every Member State. The main industrial contract normally goes to the prime contractor, which usually belongs to one of the larger Member States.

6. On earlier projects, industry organized itself in groups or consortia. This policy has gradually been changed because it introduced severe imbalances in the return coefficients. In their Phase B proposal, potential prime contractors propose a full industrial structure respecting as closely as possible the industrial return targets as established by ESA. The prime contractor's proposal is judged not only on its technical and financial merit but also on how it meets these targets.

3. The Agency and Its Member States

1. Prodex is the abbreviation of "Programme de Développement d'Expériences." It is an optional program financed by those Member States which have no built-in mechanisms to fund their space projects when they have been selected on an ESA satellite. For example, there exist cases where the total financial contribution of a Member State to space activities, both European and national, is included in the ESA contribution, and where mechanisms do not exist to redistribute money inside the country to pay for the development of experiments or payloads. The Prodex program allows this possibility. Austria, Belgium, Ireland, Norway, Switzerland, and Denmark are at present participating in Prodex. Obviously, in this program, the return coefficient should be strictly equal to 100 percent for each participating Member State, after deduction of ESA's overhead charges.

4. International Connections

1. Douglas R. Lord, "Spacelab—An International Success Story," NASA Scientific and Technical Information Division, 1987.
2. Reimar Lüst, "U.S. Cooperation in Space," *European Affairs* 3, 1989.
3. Reimar Lüst, "Cooperation between Europe and the United States in Space," *ESA Bulletin,* no. 50, 1987.
4. There exist two autonomous space agencies in Japan. ISAS, the Institute for Space and Aeronautical Sciences, is a fundamental research institution, mostly concerned with space science. NASDA, the National Space Development Agency, is mainly concerned with applications (including Earth observation) and the manned space program.

5. Two Cases of International Cooperation

1. For spacecraft traveling away from the Sun, like ISPM, the use of solar cells converting the Sun's light into electric power is not practical. Radioisotope thermoionic generators provide a more stable and compact source of energy.

2. An excellent description of the history of the early negotiations of the space station agreements can be found in the book by Joan Johnson-Freese, *Changing Patterns of International Cooperation in Space* (Orbit Book Company, 1990).

3. CRAF-Cassini, at the origin, was a combination of the Comet Rendez-Vous and Asteroid Fly-By mission (CRAF) and an orbiter of Saturn (Cassini). Both missions involved substantial international cooperation, with Germany building the propulsion system on CRAF, and ESA the Huygens probe released from the Cassini orbiter into the atmosphere of Titan for the study of its aeronomy and of its surface. In addition, Italy contributed the communication package with the probe on the orbiter. In January 1992, confronted with financial difficulties and the need to reduce the scope of the mission, NASA and DARA, the German space agency, made a joint decision to abandon CRAF. Simultaneously, Cassini itself was substantially simplified, keeping, however, the interface with Huygens untouched.

6. The New Context

1. Cluster makes up, with SOHO, the first cornerstone of the Horizon 2000 program. It consists of four identical satellites exploring the magnetosphere and the ionosphere of the Earth in three dimensions. In the framework of a cooperative agreement, the Soviets were supposed to launch a fifth satellite named Regatta-Cluster, in the 1995 time frame, so that it could add another point of measurement to the four already provided by Cluster, and increase the redundancy of the overall mission. Regatta was supposed to be developed entirely by IKI and operated jointly with ESOC. As much as possible, the instrumentation was identical, or as close as possible, to that of the Cluster payload and developed jointly between Europe and the Soviet Union.

BIBLIOGRAPHY OF WORKS IN ENGLISH

Beatle, I. J., and J. de la Cruz. "ESRO and the European Space Industry." *ESRO Bulletin,* 1967, no. 3, pp. 3–8.

Bonnet, R. M. "Towards the Selection of ESA's Next Medium Size Scientific Project (M2)." *ESA Bulletin,* 1991, no. 66, p. 37.

Bonnet, R. M. *Vanishing Horizons.* Cambridge: Cambridge University Press, 1994.

Cavallo, G. "Second Special Colloquium of the Science Programme Committee on National Space Science Programmes." ESA Science Programme Committee Document ESA/SPC(91)36, 1991.

"Convention of the European Space Agency and Rules and Procedure of ESA Council." Scientific and Technical Publication Branch of ESA.

"Delegate Bodies—Programme Boards—Terms of Reference." ESA Council Document no. 17, 1989.

Dondi, G. "The Agency's Industrial Policy: Its Principles and Their Implementation since 1975." *ESA Bulletin,* 1980, no. 21, pp. 76–83.

"The European Community—Crossroads in Space." Report by an advisory panel on the European Community and Space, Commission of the European Communities, EUR, 4010, 1991, pp. 1–44.

Johnson-Freese, J. *Changing Patterns of International Cooperation in Space.* Florida: Orbit Book Company, 1990.

Keynan, A. "The United States as a Partner in Scientific and Technological Cooperation: Some Perspectives from Across the Atlantic." Consultant Report for the Carnegie Commission on Science, Technology, and Government, 1991.

Krige, J. "The Early Activities of the COPERS and the Drafting of the ESRO Convention (1961–1962)." ESA HSR-4, January 1993, pp. 1–45.

Krige, J. "Europe into Space: The Auger Years (1959–1967)." ESA HSR-8, May 1993, pp. 1–74.

Krige, J. "The Launch of ELDO." ESA HSR-7, March 1993, pp. 1–36.

Lord, D. R. "Spacelab—An International Success Story." NASA SP 487, NASA Scientific and Technical Information Division, 1987.

De Maria, M. "Europe in Space: Edoardo Amaldi and the Inception of ESRO." ESA HSR-5, March 1993, pp. 1–36.

Marsh, D. "The ESRO Large Astronomical Satellite (LAS) Project—The Observatory in Orbit." *Journal of the British Interplanetary Society,* 1989, vol. 22, pp. 189–201.

Massey, Sir Harrie, and M. O. Robbins. *History of British Space Science.* Cambridge: Cambridge University Press, 1966.

Micklitz, H. W., and N. Reich. *Legal Aspects of European Space Activities.* Baden-Baden: Nomos Verlagsgesellschaft, 1989.

"Return to the Moon—Europe's Scientific Priorities for the Exploration and Utilisation of the Moon." ESA-SCI(91)8, 1991.

Russo, A. "The Definition of a Scientific Policy: ESRO's Satellite Programme in 1969–1973." ESA HSR-6, March 1993, pp. 1–55.

"Space Science—Horizon 2000." ESA SP 1070, 1984.

"Wishes of Industry Regarding ESA Industrial Policy and the Role of the EC in Space after 1992." Eurospace Strategic Survey, step 1, 1989.

"World Space Industry Survey—10 Years Outlook—1991–1992." Paris: Euroconsult, 1991.

ESRO AND ESA SPACECRAFT

ESRO-2B   The first ESRO satellite, devoted to the study of cosmic rays and solar X-rays. Weight: 75 kg. Telemetry data flow rate: 128 bits per second (bps).

ESRO-1A   Satellite devoted to the study of auroras and of the ionosphere. Weight: 86 kg. Telemetry data flow rate: 320bps.

HEOS-1   First highly elliptic orbiting ESRO satellite, devoted to magnetospheric research and the study of Sun-Earth relations. Weight: 108 kg. Telemetry data flow rate: 12bps.

ESRO-1B   Satellite devoted to the study of auroras and of the ionosphere. Weight: 86 kg. Telemetry data flow rate: 320bps.

HEOS-2   Satellite devoted to the study of the polar cusps, the neutral points, and the interplanetary medium. Weight: 117 kg. Telemetry data flow rate: 32bps.

TD-1   First astronomical ESRO satellite, devoted to ultraviolet, X-ray, and gamma-ray astronomy. Weight: 471 kg. Telemetry data flow rate: 1700bps.

ESRO-4   Satellite devoted to the study of the Earth's atmosphere, its ionosphere, and auroral particles. Weight: 115 kg. Telemetry data flow rate: 10240bps.

COS-B   First European gamma-ray astronomy satellite. From high elliptic orbit, COS-B could map the galaxy. It measured the diffuse emissions and observed 25 discrete sources. Its angular resolution was between 2 and 5 degrees, and it operated for nearly seven years.

GEOS-1/2   GEOS-1, a magnetospheric research satellite, was supposed to operate from a geostationary orbit. Unfortunately, a problem with the launcher placed it on an elliptical 12-hour orbit. Launched on 20 April 1977 with a Delta, its operations were terminated on 23 June 1978. It was followed by

GEOS-2, a second model, which was placed on the right orbit on 24 July 1978 and operated successfully until 1985.

ISEE-2   One of a triad of simultaneously flying satellites developed in cooperation with NASA for the study of Sun-Earth relations. It was an example of a very successful ESA-NASA cooperation. Launched on 22 October 1977, it was operated for 10 years.

IUE   The International Ultraviolet Explorer was an ESA-NASA-U.K. cooperative venture. It was the successor of the Large Astronomical Satellite of ESRO which never left the drawing board. Launched on 26 January 1978, it is still operational and is one of the greatest successes of space astronomy, proving the wealth of the UV domain for the investigation of a large number of astrophysical phenomena. It is another great success of international cooperation.

Exosat   The first ESA X-ray astronomy satellite, used as an observatory by a broad scientific community. From a highly elliptical orbit of 90h period, with a 190,000 km apogee, it could study in detail the variability of a large number of X-ray sources. EXOSAT also discovered the so-called Quasi-Periodic Objects.

Giotto   The Comet Halley mission was ESA's first deep space mission and the main component of an international effort to study Comet Halley, coordinated by the Inter Agency Consultative Group (IACG). Giotto encountered the comet on 14 March 1986 at 600 km from the nucleus. After six years of hibernation, Giotto was reactivated and directed to encounter Comet Grigg-Skjellerup on 10 July 1992 within 200 km of the nucleus. Giotto is now in hibernation once more. It will return close to the Earth in 1999. However, its reserves of gas may not allow it to be directed to a third comet.

Hipparcos   This is the first astrometric satellite in the world. It is dedicated to the precise positional measurement of some 300,000 selected stars with typical accuracies of 0.002 arcsec for each parallax and positional component, and 0.002 arcsec per year for proper motions. More than 1 million star positions to within 0.1 arcsec have also been obtained with very precise three-color photometry using the stellar sensor on board. Hipparcos was launched on 8 August 1989 with an Ariane launcher from the Centre Spatial Guyanais at Kourou. Unfortunately, its apogee motor refused to ignite and it was left in a 600 × 35,000 km orbit. In spite of this severe flaw, Hipparcos had met its scientific objectives when it stopped in 1993.

HST   The Hubble Space Telescope is the largest project in astronomy developed to date. It is a NASA project in cooperation with ESA, the latter providing the Faint Object Camera, a major scientific instrument, the solar arrays, and operational support at the Space Telescope Science Institute in Baltimore. In return, ESA receives a minimum of 15 percent of observing time. In reality, European astronomers are getting nearly 20 percent. In spite of the spherical aberration of its main mirror, the HST has observed and discovered a large number of astounding phenomena and objects. Its successful servicing mission in December 1993 has restored all its capacities, making the HST the best telescope ever built.

Ulysses Originally called Out-of-Ecliptic Mission, then International Solar Polar Mission (ISPM), Ulysses is the first project to leave the plane of the Ecliptic, after a swing-by of Jupiter (February 1992) and to fly over the poles of the Sun (south: May-September 1994, north: May-September 1995) aiming at the study of the properties of the interplanetary medium as a function of heliographic latitude. Formerly made up of two spacecraft, one European and one American, the mission now consists of only one spacecraft; the Americans decided in 1981 to abandon their own. It is still conducted cooperatively, with NASA providing the launch, the RTG, and part of the payload, and ESA the spacecraft, the operations, and the rest of the payload. Ulysses was launched on 6 October 1990. Its mission may be extended after 1995, in order to cover one complete solar activity cycle.

ISO ESA's Infrared Space Observatory is the most ambitious infrared project now under development in the world. It will operate a cryogenically cooled 60 cm telescope for the observation of astrophysical objects between 2.5 and 200 microns. Four focal plane instruments constitute the payload: one camera, two spectrometers, and one photometer. ISO will be used as an observatory open to the wide astronomical community. In the framework of an agreement with Japan and the United States, ISAS will assist ESA in implementing a third shift, and NASA will provide a second ground station, offering a complete coverage of the 24-hour orbit. The launch is planned for September 1995 with an Ariane-4 launcher.

STSP The Solar Terrestrial Science Program is the first cornerstone of Horizon 2000 and consists of two projects: Soho (the Solar and Heliospheric Observatory) and Cluster (a group of four spacecraft designed for the three-dimensional study of structures in the Earth's plasma). It is an ESA/NASA cooperative venture, under the leadership of ESA. The foreseen launch date of Soho is July 1995 with an Atlas rocket, and that of Cluster is December 1995 with the first Ariane-5 launcher. Soho will be placed in a halo orbit around the Lagrangian point L1, while the four Cluster spacecraft will operate in a polar orbit between 4 and 22 Earth radii.

Huygens This project is part of the NASA Cassini mission, which consists of an orbiter of Saturn and the ESA atmospheric probe which will land on Titan after a two-hour descent into its atmosphere. The name of the Dutch astronomer who discovered Titan in 1655 has been given to the European probe. The planned launch date is October 1997 with a Titan/Centaur rocket. The arrival at Saturn is foreseen in 2004.

XMM The X-ray Multi-Mirror Mission is a high-throughput spectroscopy observatory. It is the second cornerstone of the Horizon 2000 long-term program. It will be operated as an observatory open to the wide astronomy community, in a highly elliptical 24h-period orbit with emphasis on high sensitivity and spectral resolution. The telescope is made up of three identical units, each built from 58 individual grazing incidence mirrors. The planned launch date is 1999 with an Ariane-4.

Rosetta The mission consists of a close rendezvous with a comet and the landing of one or two surface science packages on the nucleus in order to study in situ the pristine material of the nucleus with a view to determining its structure, chemical composition, and characteristics. It is the third cornerstone of Horizon 2000. Its launch date is 2003 with Comet Wirtanen as a target. En route, Rosetta will also observe several asteroids.

FIRST The Far Infrared Space Telescope is a 3m telescope operating between 100 microns and 1 mm, using heterodyne as well as direct detection techniques. Its focal plane instruments will consist of a multifrequency heterodyne receiver and a far infrared receiver. It is the fourth cornerstone of Horizon 2000. It will be launched by Ariane in 2005.

THE HORIZON 2000 LONG-TERM PLAN

The long-term plan Space Science: Horizon 2000 (described in detail in ESA Special Publication SP-1070) was established in 1984 with wide involvement of the scientific community and was adopted by the Agency in 1985. Its realization was made possible by the commitment of ESA's Member States to provide a 5 percent annual increase in funding for the Scientific Program over about a decade.

Horizon 2000 consists of four large, six medium-sized, and possibly also a number of small projects, covering all major fields of space science. A further element of this long-term plan consists of technology studies for missions "beyond Horizon 2000."

The centerpiece of this program is a set of four so-called cornerstones which are major missions (each costing on the order of 400 MAU, at 1984 economic conditions) with scientific aims that were determined at the program's outset, covering the interests of the wide community of space researchers in Europe:

In Solar System exploration:

- The Solar Terrestrial Science Program (STSP) comprising two medium-sized missions, a solar physics mission (Soho) and a magnetospheric physics mission (Cluster), both to be launched in 1995.
- The Rosetta mission, to study pristine cometary material for detailed analysis of the early history of the Solar System.

In astronomy/astrophysics:

- The high-throughput X-ray spectroscopy mission XMM, a facility-class X-ray astrophysics observatory with an anticipated lifetime of more than ten years, to be launched in 1999.

- ■ The Far-Infrared Space Telescope (FIRST), a heterodyne and direct detection spectroscopy mission in the largely unexplored 100μm–1mm region of the electromagnetic spectrum.

In contrast to the cornerstones, for which the general scientific aims were defined at the outset of Horizon 2000, the medium-sized and small missions are selected competitively one by one. A medium-sized mission costs on the order of 200 MAU (1984 economic conditions) while a small project costs significantly less.

The first medium-sized mission selected is Huygens, a probe to be launched in October 1997 with NASA's Cassini spacecraft. Cassini will carry the Huygens probe to the vicinity of Saturn's moon Titan, where it will be released for a descent through Titan's atmosphere and a landing on its surface in 2004.

Candidates for future medium-sized missions will be studied in response to ESA's calls for mission proposals. The selection procedure for medium-sized missions has three steps:

1. Selecting from the missions proposed by the scientific community, up to six missions for a study at assessment level.
2. Selecting from the missions under the study at assessment level up to four missions for a study at Phase A level.
3. Selecting from the missions under study at Phase A level one mission to be carried out.

The selections 1 and 2 are made by the Space Science Advisory Committee (SSAC) on the basis of recommendations from the Astronomy Working Group for astronomy/astrophysics missions, the Solar System Working Group for Solar System exploration missions, and an ad hoc advisory panel for missions in fundamental physics.

The number of candidate missions is reduced from one selection to the next and the remaining missions are studied in greater depth. The objective of the two study phases is to define the mission science objectives, the scientific model payload, the spacecraft, launcher, operations, and, where applicable, the share of tasks between collaborating space agencies, to a level where a schedule and cost-at-completion can be estimated with some confidence.

At assessment level, a study typically lasts six months, followed by a consolidation phase of three months. At Phase A, a study typically lasts twelve months, followed again by a consolidation phase of three months. The consolidation phase is needed for performing cost and technical reviews, identifying cost drivers and potential reductions in mission scope, and fine-tuning international collaborative plans.

All technical work at assessment level is carried out in-house; at Phase A level, noncompetitive technical studies are done by industry under ESA contract. During both study phases ESA is advised on scientific matters by a study team composed of external scientists, selected by ESA based on their competence in the particular discipline of the mission under study.

Through the strong involvement of the scientific community at all study phases, and in all steps of the selection process, and by coupling the competence of external scientists with in-house scientific and technical expertise, it is possible to select ultimately the single best mission from the number proposed at the outset.

For the more distant future, "beyond Horizon 2000," ESA is carrying out technological studies for the following possible missions:

- A solar probe, Vulcan, that would approach the Sun to within four solar radii.
- Elements of a mission to Mars, in particular a network of hard and semi-hard landers, a rover, and a Mars orbiter.
- Astronomical imaging through two-dimensional interferometry from space.
- Opportunities that might be offered by a lunar base for science "from the Moon," "on the Moon" and science "of the Moon."

These studies do not, at this stage, represent a commitment on the side of ESA to create new cornerstones in the relevant scientific fields.

LIST OF ILLUSTRATIONS

Figure 1. Evolution of the annual budgets of ESRO, ELDO, and ESA 1962–1991   7

Figure 2. Ariane on its launch pad at Kourou   15

Figure 3. The International Ultraviolet Explorer (IUE)   18

Figure 4. ESA's Giotto mission to Halley's Comet   19

Figure 5. Flow chart for the planning and implementation of ESA's optional programs   31

Figure 6. The structure of the Horizon 2000 program   40

Figure 7. The structure of ESA's delegate bodies   42

Figure 8. The succession and timing of the various phases of development of a project   52

Figure 9. Flow chart showing the interplay between ESA, the delegations, and industry during pre-Phase B and Phase B for scientific projects   54

Figure 10. The diversity of European organizations responsible for space   67

Figure 11. Comparison, in millions of dollars, of the most significant space programs, 1990   68

Figure 12. A picture of Northern Holland taken by ERS-1   85

Figure 13. Spacecraft involved in the space missions to Halley's Comet   89

Figure 14. The Pathfinder Concept   91

Figure 15. The ceremony at the Vatican on 7 November 1986, when the heads of the IACG delegations presented the results of the missions to Halley's Comet to Pope John Paul II   93

Figure 16. The Solar Terrestrial Science Program missions which are being coordinated by the IACG   95

Figure 17. The present scenario of Ulysses   107

Figure 18. The originally planned European contribution to the international space station   110

Figure 19. The NASA Cassini mission: an orbiter of Saturn to study the planet, its rings, and its satellites; and the Huygens probe to study Titan   113

INDEX

Advanced X-ray Astrophysical Facility (AXAF; NASA), 77, 112
Aerospace industry. *See* Industry
AFC. *See* European Space Agency: Administrative and Finance Committee of
Afghanistan, 82, 92
Africa, 87
Air and Space Museum (Washington, D.C.), 81
Albee, Arden, 80
Amaldi, Edoardo, 4
Announcements of Opportunity, 80, 87
APM. *See* Attached Pressurized Module
Applications programs, 4, 17, 20, *22,* 24–25, 30, 60, 64, 73, 128, 136; of India, 86; of Japan, 143n4; military, 111, 122
Aranzadi, Claudio, 130
Argentina, 86–87
Ariane launcher, 4, 11, 14, *15,* 20, 59, 64, 79, 108, 148, 150; Ariane-1, *19;* Ariane-4, 142n1, 149, 150; Ariane-5, 11, 129; commercial use of, 120, 127–128, 136, 137; and ESA/NASA relations, 77; and Optional Programs of ESA, 29, 44
Arianespace, 137
Armstrong, Neil, 121
Asca satellite (Japan), 84
Asteroid Fly-By. *See* CRAF mission
Astrometry, 43, 142n1, 143n4, 148
Astronauts, 83, 122. *See also* Manned space flight; *individual astronauts*

Astronomy, 43, 46, 76, 136, 147, 149, 150, 151, 152, 153; gamma ray, 83, 122, 147; optical, 86; ultraviolet, 16, 86–87, 99, 147; X-ray, 84, 99, 122, 147, 148, 151. *See also individual missions*
Astrophysics, 37, 38, 59, 94, 105, 122, 148, 149, 151, 152
Atlas rocket, 149
Attached Pressurized Module (APM), 109, *110,* 111, 116. *See also* Columbus program; Space station
Auger, Pierre, 4, 5, 9
Auroras, 16, 147
Australia, 12, 86–87
Austria, 10, 78, 111, 124, 143n1
AXAF. *See* Advanced X-ray Astrophysical Facility

Bad Godesberg Space Conference, 10, 17
Bannier, Jan, 17
Beggs, James, 105, 106, 108
Belgium, 4, 10, 12, 16, 53, 78, 132, 142n9, 143n1
Biermann, Ludwig, 16
BION-10 spacecraft (Russia), 83
Bleeker, Johan, 37, 38
Blue Streak (U.K.), 11, 13. *See also* Europa Launcher
BNSC. *See* British National Space Centre
Bolin, Bert, 142n5
Bondone, Giotto di, 88
Bonnet, Roger M., 37, 41, 80, 94, 114

Brazil, 87
Britain. *See* United Kingdom
British National Space Centre (BNSC), 65–66
Broglio, Luigi, 5, 8
Brussels (Belgium), 11, 20, 125
Bulgaria, 83
Buran (Soviet space shuttle), 121, 122
Bush, George, 114

Canada, 86–87, 108, 114, 115, 116, 117, 118, 138, 142n9
Canadian Space Agency, 109
Capri (Italy), SPC meetings in, 28, 69–70, 71, 72
Carrobio di Carrobio, Renzo, 13
Carter, Jimmy, 102, 115
Cassini mission, *113,* 115, 144n3, 150, 152
Causse, Jean-Pierre, 17
Cavallo, Giacomo, 69
Center for Space Science and Applied Research (CSSAR; China), 87
Central America, 87
Central Europe, 81–84, 123–124
Centre National d'Etudes Spatiales (CNES; France), 14, 57, 64–65, 132, 142n7
Centre Spatial Guyanais, 142n7, 148
CERN. *See* European Center for Nuclear Research
CESAR satellite, 124
Challenger. *See* Space Shuttle: Challenger
Chile, 87
China, 66, 87, 120, 135, 137, 142n2
Chinese Academy of Sciences, 87
Clarke, Kenneth, 132
Climate, 39, 128
Clinton, Bill, 115, 116
Cluster mission, 76, 83, 87, *95,* 122, 144n1, 149, 151
CNES. *See* Centre National d'Etudes Spatiales
Cold war, 11, 92, 117, 122, 132
Colombo, Giuseppe, 88
Colombia, 87
Columbus program, 109–112, 116, 129, 130, 131, 132. *See also* Space station
Comet, Stig, 9
Comet missions, 16, 17, 23. *See also*

CRAF mission; Giotto mission; Rosetta mission
Comet Rendez-Vous mission. *See* CRAF mission
Committee for Space Research (COSPAR), 141n1
Coordination Committee on Multilateral Export Control, 82
Coralie, 12. *See also* Europa launcher
COS-B satellite, 30, 75, 100, 147
Cosmic rays, 14, 147
Cosmonauts. *See* Astronauts
COSPAR. *See* Committee for Space Research
CRAF mission (Comet Rendez-Vous, Asteroid Fly-By), 112, 114, 115, 144n3
Crowley, Thomas, 9
CSSAR. *See* Center for Space Science and Applied Research
Curien, Hubert, 65, 108, 132
Czechoslovakia, 83, 124

D-1 mission (Germany), 79
D-2 mission (Germany), 79
d'Allest, Frederic, 65
DARA (German Space Agency), 144n3
Data, dissemination of, 63, 87, 96, 97, 103, 123, 137
Deep Space Network (DSN), 90, *91*
de Gaulle, Charles, 12, 64
de Jager, Kees, 46, 80
Delehais, Maurice, 38
Delft Technical University, 8
Delors, Jacques, 125
Delta launcher, 147
Denmark, 10, 12, 17, 20, 78, 142n9
Discovery. *See* Space Shuttle: Discovery
Donahue, Thomas, 80
Dordain, Jean-Jacques, 131
Dornier (German company), 104, 105
DSN. *See* Deep Space Network
Dupree, Andrea, 80

Earth observation, 39, 41, 128, 130, 131, 133, 136; and Australia, 86; and China, 87; cooperation in, 78, 95, 123; and EEC, 125; and India, 86, 87; and Japan,

143n4; and space station, 109, *110;* and USSR, 83, 122

Earth Observing System (U.S.), 109

Earth science, 36, 39, 41, 78, 79, 123, 128

Ecliptic plane, 98, 106, 149

Ecuador, 87

Edelson, Burt, 94

EEC. *See* European Economic Community

ELDO. *See* European Launcher Development Organization

Energia launcher (Russia), 122

Engineers, 112, 117, 123, 124

England. *See* United Kingdom

Environment, 39, 112, 122, 125, 127, 128, 129, 130, 133

Envisat-1 satellite, 133. *See also* Columbus program

ERS-1 satellite, *85,* 86, 87, 131, 136

ESA. *See* European Space Agency

ESC. *See* European Space Conference

ESDAC. *See* European Space Data Center

ESLAB. *See* European Space Laboratory

ESOC. *See* European Space Operation Center

ESRANGE (launching base), 5, 9, 14, 17

ESRIN. *See* European Space Research Institute

ESRO. *See* European Space Research Organization

ESRO-1A satellite, 16, 75, 147

ESRO-1B satellite, 75, 147

ESRO-2B satellite, 14, 75, 147

ESRO-4 satellite, 20, 75, 147

ESTEC. *See* European Space Technology Center

ESTRACK (tracking network), 5

Eumetsat company, 137

Europa launcher, 12, 13–14, 16, 20, 79

European Center for Nuclear Research (CERN), 4, 6, 37, 58–59

European Commission, 120, 124, 125, 127, 128. *See also* European Economic Community

European Conference for Space Telecommunications, 10

European Currency Unit (ECU), 141n2

European Economic Community (EEC), 56, 120, 124–128, 141n2

European Environmental Agency, 128

European Launcher Development Organization (ELDO), 5, *7,* 8, 10, 11–14, 22, 64, 120, 142n7

European Science Foundation, 37

European Southern Observatory, 37

European Space Agency (ESA): Administrative and Finance Committee of (AFC), 41, 44, 100, 142n6; approval procedures of, 99–100; Astronomy Working Group of, 33, 34–35, 38, 43, 69–70, 152; budget of, 6, *7,* 27, 28–29, 36–37, 39, 47, 50, 58, 63, *68,* 70, 75, 98, 100–101, 118, 128–129, 138; Convention of, 20, 24, 30, 41, 45, 50, 60, 64, 124, 125, 126; Council of, 24, 28, 30, 38, 39, 41, 44, 45, 48, 57, 65, 66, 71, 82, 100, 101, 108, 109, 112, 125, 129, 130, 132, 141n2; Executive of, 24, 33, 34, 36, 41, 43, 44, 45, 50, 51, 53, 76, 80, 83, 103, 105, 129, 130, 131, 132; Future Program Study Office of, 38; Industrial Policy Committee of, 41, 44, 51, 53; industrial policy of, 47–*52,* 53–*54, 55*–57, 66, 120, 125, 126, 137; Management Board of, 104; Mandatory Program of, 25–29, 32, 39, 43–44, 65, 70, 137–138, 142n3; Optional Program of, 10, 29–30, *31,* 32, 36, 44–45, 64, 137–138; Program Boards of, 44, 45; Science Projects Department of, 38; Solar System Working Group of, 33, 34–35, 38, 43, 69–70, 152; Space Science Advisory Committee (SSAC) of, 33–34, 37, 43, 69–70, 77, 99, 152; Space Science Department (SSD) of, 38, 58, 82, 100; structure of, *42;* Survey Committee of, 36, 37, 38; Washington office of, 44, 75, 114, 134. *See also* Science Program; Science Program Committee

European Space Conference (ESC), 10, 11, 13, 17, 20, 23

European Space Data Center (ESDAC), 5, 8, 9

European Space Laboratory (ESLAB), 5, 8, 9, 58, 141n4

European Space Operation Center (ESOC), 8, 90, 144n1

European Space Research Institute (ESRIN), 9, 10, 17

European Space Research Organization (ESRO), 3–10, 32, 36, 59, 60, 95, 120, 136, 137, 142n5; Blue Book of, 5, 6; budget of, 6–8, 11, 16, 50, 141n2; Convention of, 9–10, 17, 22, 30, 48, 62; Council of, 48; and ELDO, 11–14, 16, 17, 20, 22; industrial policy of, 47–48; and NASA, 57–58, 75, 77, 99; and national space organizations, 62, 64; Seat of, 5, 8, 80, 108, 132, 141n3; spacecraft of, 14, 16, 20, 75, 147–150; Space Science Department (SSD) of, 141n4; and USSR, 82

European Space Technology Center (ESTEC), 5, 8, 9, 57, 58, 141n3, 141n4

Eurospace, 126

Eutelsat company, 137

Exosat satellite, 26, 75, 84, 148

Faint Object Camera (FOC), 58, 76, 148

Fair return principle. *See* Industrial return principle

Far Infrared Space Telescope (FIRST), 53, 77, 78, 150, 152

Finland, 124, 138, 142n9

FIRST. *See* Far Infrared Space Telescope

Fisk, Lennard, 115

FOC. *See* Faint Object Camera

Focal plane instruments, 77, 149, 150

FR-1 satellite (France), 62

France, 4, 5, 10, 36, 44–46, 60, 62, 66, 122, 142n9; and applications programs, 17, 20, 64; budget of, 53, 131, 132, 142n8; and CNES, 14, 57, 64–65; and ELDO, 11, 12, 13; and the environment, 128; and 1991 Munich ESA meeting, 129–130; and Spacelab, 78; and space station, 117

French Guyana, 13, 15, 16, 142n7, 148

Gagarin, Yuri, 121

Galileo, 105

Gamma-rays. *See* Astronomy: gamma ray

Geiss, Johannes, 80

GEOS missions, 30, 75, 147–148

Geotail satellite (Japan), *95*

Germany, 4, 5, 10, 16, 45–46, 62, 66, 78, 105, 128, 142n9; budget of, 53, 142n8; and CRAF-Cassini, 114, 144n3; and 1991 Munich ESA meeting, 129–130; and 1992 Granada ESA meeting, 131–132; participation in ELDO of, 12; reunification of, 129; and space station, 109, 117

Gibbons, John H., 116

Gibson, Roy, 20, 34, 124–125, 127

Ginga satellite (Japan), 84

Giotto mission, *17, 19,* 23, 26, 32, 43, 136, 148; and IACG, 88, *89,* 90, *91,* 92; and NASA, 75, 76, 77, 108

glasnost, 82, 121

Glavkosmos (USSR), 83

Goddard Space Flight Center (NASA), 99

Goldin, Daniel, 115, 116

Gorbachev, Mikhail, 82

Granada (Spain), ESA meeting in, 11, 30, 41, 51, 110, 123, 131–133, 135, 138

Greece, 124

Grigg-Skjellerup Comet, *17,* 136, 148

Ground stations, 12, 50, 149

Hague, The (The Netherlands), ESA meeting in, 11, 50, 109, 111, 128, 129, 130, 132

Haig, Alexander, 104

Halley's Comet, *17,* 19, 32, 82, 84, 88–*89,* 90–91, 92–93, 94, 136, 148

Hanin, Charles, 20

Harwit, Martin, 81

Haskell, George, 38

Heliosphere research, 59, 98, 149

HEOS missions, 16, 20, 75, 147

Hermes European space plane, 11, 44, 65, 84, 109, *110,* 129, 130, 131, 132, 133

Hipparcos mission, 43, 58, 74, 76, 142n1, 148

Hocker, Alexander, 10

Holland. *See* Netherlands, The

Hope project (Japan), 86

Horizon 2000 program, 11, 28, 33, 36–41, 43, 63, 138, 151–153; and ESA budget, 29, 53, 66; and international cooperation, 73–74, 77, 78, 81, 96, 128; and

national space programs, 60, 69, 71–72; and Rosetta mission, 150, 151; structure of, *40;* and STSP, 144n1, 149
Hubble Space Telescope (HST), 26, 58, 70, 74, 75, 76, 77, 78, 99, 105, 148–149
Huber, Martin, 58
Hungary, 83, 124
Huygens probe, 112, *113,* 115, 144n3, 150, 152. *See also* Cassini mission

IACG. *See* Inter Agency Consultative Group
ICE. *See* International Cometary Explorer
IGA. *See* Intergovernmental Agreement
IKI. *See* Space Research Institute
India, 86, 87
Indian Space Research Organization (ISRO), 86
Industrial return principle, 49–51, 53–57, 59, 125–127, 133, 143nn6,1
Industry, 3, 47–57, 63, 79, 84, 100, 103, 116, 117, 123, 125–127; contracts in, 13, 48–51, *52, 54–55,* 56, 126–127, 142n4, 143nn5,6, 153; and market competition, 50, 56, 120, 125–128, 136, 137, 142n2. *See also* European Space Agency: industrial policy of; *individual nations*
Infrared Space Observatory (ISO), 26, 74, 75, 77, 84, 108, 149; Science Team of (IST), 81
Inmarsat, 137
In-orbit infrastructure, 83, 129
Institute for Space and Aeronautical Sciences (ISAS; Japan), 70, 82, 84, 85, 88, *95,* 143n4, 149
Integral project, 83
Inter Agency Consultative Group (IACG), 73, 82, 84, 88–*93,* 94, *95,* 97, 134, 148
Interball mission (Russia), *95*
Intercosmos (USSR), 83
Interferometry, 43, 86, 87, 95, 153
Intergovernmental Agreement (IGA), 79, 109, 111, 114, 116, 118
International Astronomical Union, 37
International Cometary Explorer (ICE; U.S.), 88, *89*
International Halley Watch, 88, 90

International Microgravity Laboratory, 78
International Solar Polar Mission (ISPM), 75, 76, 80, 98–108, 109, 115, 117, 118, 143n1. *See also* Ulysses mission
International Space Year, 134
International Ultraviolet Explorer (IUE; formerly Large Astronomical Satellite), *16,* 17, 18, 30, 32, 46, 75, 148
Interplanetary medium, 98, 147, 149
Ionosphere, 16, 144n1, 147
Ireland, 142n9, 143n1
ISAS. *See* Institute for Space and Aeronautical Sciences
ISEE mission, 75, 88, 148
ISO. *See* Infrared Space Observatory
ISPM. *See* International Solar Polar Mission
ISRO. *See* Indian Space Research Organization
Italy, 4, 5, 10, 69, 78, 124, 128, 142n9, 144n3; budget of, 53, 132; Frascati, 8; Isola di San Giorgio Maggiore, 38; Naples, 28, 69; national space program of, 60, 62; Padua, 88, 94; Rome, 7; Venice, 38. *See also* Rome, ESA meeting in
IUE. *See* International Ultraviolet Explorer
IVS mission, 43

Japan, 66, 72, 73, 75, 79, 84–86, 134, 135, 149; budget of, 70, 84; and IACG, 90, 92, 94; and industrial competition, 120, 127, 141n2; and the space station, 108, 114, 115, 116, 117, 118; Tokyo, 86. *See also* Institute for Space and Aeronautical Sciences; National Space Development Agency
Jet Propulsion Laboratory, 115
John Paul II, Pope, 92, *93,* 94
Jordan, Hermann, 9
Jupiter, 16, 99, 149
juste retour. See Industrial return principle

Kenya, 87
Kiruna (Sweden), 9, 14
Koptev, Yuri, 83
Kourou (French Guyana), 13, 15, 142n7, 148

Lagrangian Point L1, 149
Laika, 121
Language barriers, 45–47
Large Astronomical Satellite (LAS), 16, 17, 18, 23, 148. *See also* International Ultraviolet Explorer
Launchers, 4, 17, 30, 32, 50, 58, 73, 74, 86, 104; and industrial competition, 48, 120, 127–128, 136, 142n2. *See also* European Launcher Development Organization; *individual launchers*
Lawrence, T. E., 33
Leiden, University of, 8
Levi, Jean-Daniel, 132
Life sciences, 78, 83, 123
Lines, Freddy, 9
Lord, Douglas R., 79
Lovelace, Alan, 101
Lunar Polar Orbiter, 43
Lüst, Reimar, 9, 36, 79, 80, 125, 130, 142n5
Luton, Jean-Marie, 41, 83, 114, 116, 125
Lyman program, 87

Maastricht Treaty, 124, 129
Machetto, D., 76
MacMillan, Harold, 12
Magnetospheric research, 16, 144n1, 147, 151
Manned space flight, 78, 83, 87, 110, 122, 123, 130, 135, 143n4. See also *individual missions*
Manno, Vittorio, 69, 76, 96, 101, 103, 114
Man-Tended Free Flyer (MTFF), 109, *110,* 111, 132, 133. *See also* Columbus program; Space station
Mars, 122, 135, 153
Mars-94 mission (Russia), 83, 122
Massey, Sir Harrie, 4, 5, 10
Max Planck Institute for Extraterrestrial Physics, 9
McDonald, Frank, 80
Memorandum of Understanding (MOU), 75, 76, 109; for International Solar Polar Mission, 99, 100–101, 102–104, 106, 118; for Spacelab, 79; for space station, 109, 111, 114, 116
Meteorology, 29, 60, 64, 123, 133, 136, 137

Meteosat satellite, 64, 136, 137
Metop-1 satellite, 133
Mexico, 87
Microgravity, 36, 39, 41, 70, 83, 110, 136
Mir (Soviet space station), 83, 117, 122, 123
Missiles, 11, 12
Moon, 16, 79, 135, 153
Morocco, 87
MOU. *See* Memorandum of Understanding
MTFF. *See* Man-Tended Free Flyer
Munich (Germany), ESA meeting in, 11, 51, 83, 110, 112, 123, 124, 128–131, 133, 138
Mussard, Jean, 9

NASA. *See* National Aeronautics and Space Administration
NASDA. *See* National Space Development Agency
National Aeronautics and Space Administration (NASA; U.S.), 25, 44, 57–58, 72, 73, 75–82, 98, 125, 144n1. *See also* Inter Agency Consultative Group; International Solar Polar Mission; Space station; *individual missions*
National Space Development Agency (NASDA; Japan), 109, 143n4
NATO. *See* North Atlantic Treaty Organization
Ness, Norman, 80
Netherlands, The, 4, 8, 10, 12, 17, 37, 62, 78, 141nn1,4, 142n9. *See also* Hague, The
New York (U.S.), 101
NOAA satellite, 136
North Atlantic Treaty Organization (NATO), 3, 141n1
Norway, 8, 10, 124, 142n9, 143n1a
NPO Energia company (USSR), 83

Occhialini, Giuseppe, 25, 76
Ockels, Wubbo, 57
Office d'Etudes et de Recherches Aeronautiques (France), 131
Olthof, Henk, 38
Orbits, 14, 15, 88, 98, 99, 117, 120, 142n1, 147, 149, 150

Out-of-Ecliptic Mission, 99, 149. *See also* International Solar Polar Mission; Ulysses mission

Page, Edgar, 38
Panetta, Leon, 115
Paris (France), 8, 9, 13, 44, 101, 131, 132, 141n3
Pathfinder Concept, 90, *91,* 92
Payloads, 25, 32, 35, 36, 43, 84, 120, 143n1, 152
Pellat, Rene, 36
perestroika, 82
Peterson, Larry, 80
Phobos mission, 92
Photometry, 148, 149
Physics, 33, 82, 123; fluid, 78; fundamental, 33; geo-, 136; high energy, 58; magnetospheric, 151; particle, 4; plasma, 82, 88, 149
Pinkau, Klaus, 66
Pioneer mission (U.S.), 98
Planetary research, 43, 59, 96
Poland, 83, 124
Polar platform, 78, 109, *110,* 111, 133. *See also* Columbus program; space station
Polar Satellite (U.S.), *95*
Portugal, 124
Pounds, Kenneth, 80
PRISMA satellite, 43
Prodex (Programme de Developpement d'Experiences), 63, 84, 143n1
Project development, *52*
Proton launcher (Russia), 127
Pryke, Ian, 114, 134
Puppi, Giampietro, 11, 17

Quasat satellite, 86, 87
Quasi-Periodic Objects, 148
Quayle, Dan, 114
Quilles, Paul, 130, 132
Quitsgaard, Erik, 37, 38, 101, 103, 106, 108

Radioastron mission (Russia), 86, 94
Radioastronomy, 86
Radioisotope thermoionic generator (RTG), 103, 143n1

Reagan, Ronald, 101, 102, 108, 111, 115
Reciprocity Agreement, 80–81
Regatta satellite, 122, 144n1. *See also* Cluster mission
Remote sensing, 60, 122, 125, 127
Research Institute for Particle and Nuclear Physics (Hungary), 83
Riesenhuber, Heinz, 132
Roman, Nancy, 76
Romania, 83
Rome, ESA meeting in, 11, 17, 38, 39, 50, 69, 128
Rome Treaty, 125, 126
Rosetta mission, 53, 77, 150, 151
Royal Aircraft Establishment (U.K.), 9
Russia, 70, 81–83, 95, 117, 120, 121–123, 130, 134, 135, 136; and industrial competition, 127, 131, 132, 133, 137, 142n2; Space Agency of, 83, 122. *See also* Union of Soviet Socialist Republics

Sacotte, Daniel, 132
Sagdeev, Roald, 92, 97, 121, 122
Sakigake mission (Japan), *89, 90*
San Aranjuez, General, 46
San Marco-1 satellite (Italy), 62
Satellites, *21, 22, 95, 147–150. See also* Applications programs; Earth observation; *individual missions*
Saturn, *113,* 142, 144n3, 150
Schwassmann-Wachmann Comet, 150. *See also* Rosetta mission
Science Program (ESA), 30, 32, 34, 36, 41, 73, 76, 77, 78, 85–87, 112, 141n3; budget of, 28–29, 39, 53, 63, 74, 100–101, 128–129, 151; and ISPM, 99, 100–101, 103; and Mandatory Programs, 25, 26, 34, 137; and Member States, 62, 66
Science Program Committee (SPC; ESA), 25–26, 28, 41, 46, 69, 77, 83, 97, 100, 108, 112, 142n6; and Mandatory Program, 43, 44; and Member States, 28, 62, 65, 66; and SSAC, 33, 34
Single European Act, 56, 120, 124, 125, 126, 127. *See also* European Economic Community
SIRTF. *See* Space Infrared Telescope Facility

SLED equipment, 57
SMIMM project (U.S.), 78
Solar-A mission (Japan), *95*
Solar and Heliospheric Observatory
   (Soho), 77, *95,* 144n1, 149, 151
Solar arrays, 76, 148
Solar research, 83, 84, 98, 99, 147, 151.
   See also STSP; *individual missions*
Solar System exploration, 37–38, 59, 94,
   151, 152
Solar Terrestrial Science Program (STSP),
   53, 77, 78, *95,* 149, 151
Sounding rockets, 5, 6, 9, 12, 14, 17, 19,
   62, 88
South America, 87
Soviet Union. *See* Russia; Union of Soviet
   Socialist Republics
Space Act, 98
Space agencies, national, *67, 68*
Space Agency Forum, 134
Spacecraft, *21, 22,* 143n1, 147–150. See
   also *individual missions; individual
   nations*
Space Infrared Telescope Facility (SIRTF;
   U.S.), 81, 108, 112
Spacelab, 4, 20, 29, 57, 78–80, 136
Space planes, 83, 86. *See also* Buran;
   Hermes European space plane; Hope
   project; Space Shuttle
Space Research Institute (IKI; USSR), 82,
   88, 92, 121, 122, 144n1
Space Shuttle (U.S.), 20, 79, 99, 102, 104,
   109, *110;* Challenger, 100, 106, 111,
   121; Discovery, 106
Space station, international, 11, 79, 86, 98,
   108–119, 129, 135, 144n2
Spain, 16, 46, 53, 62, 78, 83, 130, 131,
   132, 142n9
Spectroscopy, 149, 150, 151, 152
Spectrum X-Gamma observatory (USSR),
   122
Sputnik-1 (USSR), 4, 121
SSAC. *See* European Space Agency: Space
   Science Advisory Committee of
Steinberg, Jean-Louis, 80
STEP project, 33
Stockman, David, 102, 104
Strategic Defense Initiative, 94

STSP. *See* Solar Terrestrial Science Pro-
   gram
Suisei mission (Japan), *89, 90*
Sun, 88, 99, *107,* 143n1, 149, 153
Sun-Earth relations, 16, 84, *95,* 147, 148
Superconducting Super Collider (SSC;
   U.S.), 118
Sweden, 4, 8, 9, 10, 12, 14, 17, 62, 111,
   124, 142n5, 142n9
Swings, Pol, 16
Switzerland, 4, 8, 10, 12, 58, 59, 78, 111,
   142n9, 143n1

TD-1 satellite (European Space Research
   Organization), 16, 20, 75, 147
Telecommunications, 16, 29, 60, 70, 85,
   86, 87, 123, 127, 137
Telemetry, 5, 12
Temple 2/Halley mission, 76
Thorneycroft, Sir Peter, 12
Titan, *113,* 144n3, 150, 152
Titan/Centaur rocket, 150
Tokyo (Japan), 86
Tracking stations, 4, 14, 58
Trendelenburg, Ernst, 9, 92, 103, 141n3
Truly, Richard, 114
Tunisia, 87

UK-1 satellite (U.K.), 60
Ultraviolet astronomy. *See* Astronomy:
   ultraviolet
Ulysses mission (formerly ISPM), 70, 76–
   77, 106, *107,* 118, 149
Union of Soviet Socialist Republics
   (USSR), 3, 4, 73, 81–84, 121–123, 129;
   and cooperation with individual nations,
   62, 65, 66, 72; and IACG, 84, 88, 90.
   See also Russia; Space Research Institute
United Kingdom, 45, 57, 62, 78, 84, 100,
   111, 141n3, 142n9; budget of, 6, 39, 53,
   65–66, 132, 142n8; and ELDO, 11, 12,
   13; and ESRO, 4, 5, 6, 8, 9, 10; and
   International Ultraviolet Explorer, 17,
   46, 148
United States, 3, 4, 11, 14, 23, 79, 86, 123,
   124, 127, 133, 136; budget of, 70, 75,

98, 101–106, 106, 111, 112, 114–118; Congress of, 99, 104, 105, 108, 112, 114, 115, 117; and cooperation with individual nations, 65–66; and IACG, 92, 95; and industrial competition, 49, 137, 142n2; and ISPM, 98–108; Office of Management and Budget (OMB) of, 102, 105, 109, 115, 117; and reciprocity, 80–81; and space station, 108–119; State Department of, 104, 105, 107. *See also* National Aeronautics and Space Administration
USSR. *See* Union of Soviet Socialist Republics

Valente, Saverio, 28, 69
Van Allen belts, 142n1
Van de Hulst, Henk, 5, 10, 141
Vatican, 92, *93*

Vega missions (USSR), *89,* 90, *91,* 92
VSOP mission (Japan), 94
Vulcan probe, 153

Washington, D.C. (U.S.), 20, 111
Weather forecasting, 136, 137
Whitcomb, Gordon, 38
Wind Satellite (U.S.), *95*
World Space Agency, 133
World War II, 3, 4

X-ray Multi-Mirror Mission (XMM), 53, 77, 78, 81, 150, 151
X-rays, 14, 147. *See also* Astronomy: X-ray

Yohkoh. *See* Solar-A mission

Zimbabwe, 87